The Balearic Pentapartite Division of Prehistory

Radiocarbon and other age determination inventories

William H. Waldren

BAR International Series 282
1986

B.A.R.

122 Banbury Rd, Oxford, OX2 7BP, UK

B.A.R.-S282, 1986: 'The Balearic Pentapartite Division of Prehistory'

ISBN 9780860543640 paperback
ISBN 9781407344577 e-book
DOI https://doi.org/10.30861/9780860543640
A catalogue record for this book is available from the British Library

This book is available at www.barpublishing.com

Acknowledgements

This publication is dedicated to a very dear friend, **Walter O. Heinze**, who first supported these investigations as early as 1962 and still contributes at present.

Appreciation is extended to the many academic and scientific institutions and their facilities that have carried out the different analyses, free of any charge. These are: the **University of Washington at Seattle, Yale University, Smithsonian Institution of Natural History, Scripps Institute of Oceanography, University of California at Los Angeles, British Museum, Harwell Laboratories, Oxford University and the Spanish Centro Superior de Investigaciones Cientificos**.

Special thanks and the kindest of regards are offered my very good friend, **Minze Stuiver**, who has been responsible for the bulk of the datings in this inventory.

Further appreciation is given to another friend, **Richard Burleigh**, who also has given assistance.

CONTENTS

II

PRIMARY, SECONDARY AND TERTIARY RESEARCH SITES

<<<<<<<<<<<<<<<<<<<<<<<<<<<<<<<<<<<<<<<<<<<<>>>

III

PENTAPARTITE DIVISION OF BALEARIC PREHISTORY

II. The Early Settlement Period

III. The Pretalayotic Period

IV. The Talayotic Period

V. The Post Talayotic Period

IV

OBSERVATIONS, CONCLUSIONS AND BIBLIOGRAPHY

‹‹‹‹‹‹‹‹‹‹‹‹‹‹‹‹‹‹‹‹‹‹‹‹‹‹‹‹‹‹‹‹‹‹‹››››››››››››››››››››››››››››››››››››

V

AGE DETERMINATION AND OTHER INVENTORIES

Figures and Tables

Figures:

Tables:

Please note that the foldout that was originally available at the end of this book is now available to download from www.barpublishing.com/additional-downloads.html.
The original foldout has been reduced in size to match the A4 format of this book, the image is therefore not as clear as the original foldout. Please refer to the original foldout via the download for the original content.

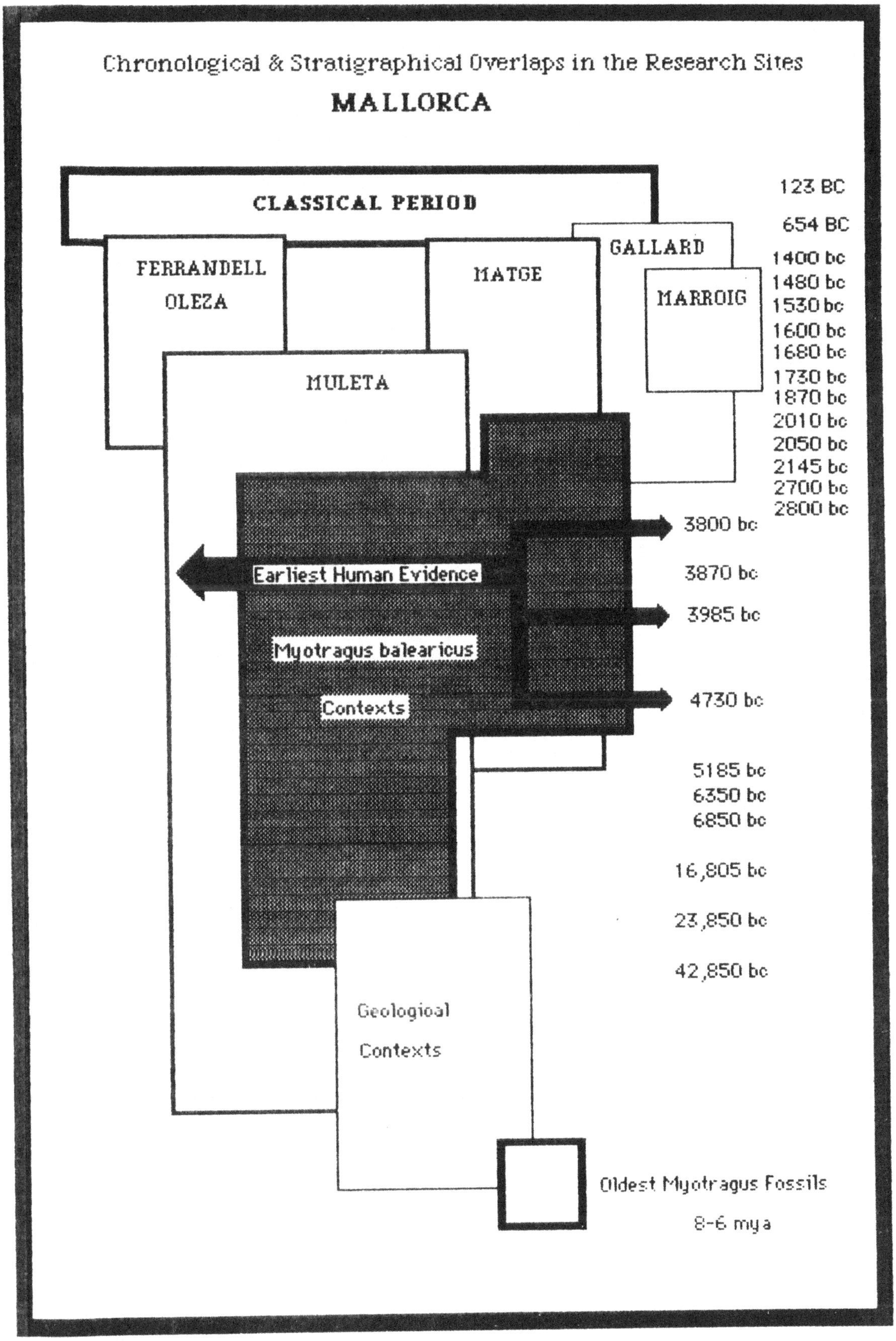
Chronological & Stratigraphical Overlaps in the Research Sites
MALLORCA
CLASSICAL PERIOD
FERRANDELL
OLEZA
MATGE
GALLARD
MARROIG
MULETA
Earliest Human Evidence
Myotragus balearicus
Contexts
Geological
Contexts
Oldest Myotragus Fossils
8-6 mya
123 BC
654 BC
1400 bc
1480 bc
1530 bc
1600 bc
1680 bc
1730 bc
1870 bc
2010 bc
2050 bc
2145 bc
2700 bc
2800 bc
3800 bc
3870 bc
3985 bc
4730 bc
5185 bc
6350 bc
6850 bc
16,805 bc
23,850 bc
42,850 bc

FIGURE 1

Preface

The present publication has been compiled around an earlier nucleus of absolute dating analyses results for Balearic prehistoric contexts compiled and first published in 1981. This early listing, included as an appendix, formed part of the author's published University of Oxford Doctor of Philosophy Dissertation, '**Balearic Prehistoric Ecology and Culture.....the Excavation of Certain Caves, Rock Shelters and Settlements** (**Valdren 1982**).

The inventory list at the time of its first publication consisted of 100 radiocarbon and other chronometric analyses available for the Balearic prehistoric sequence (**Valdren 1981**). Up until then, several interim reports and still earlier partial lists of such dates had been published in Spanish scientific journals and other site reports (**e.g. Rossello Bordoy, Kopper and Valdren 1967 and Fernandez-Miranda and Valdren 1974**). However, this current inventory, recompiled and significantly added to since the original 1981 edition, offers the reader the most up-to-date and comprehensive listing of chronometric dates; representing as it does at present the results of 24 years compilation by the author since 1962.

As well as a comprehensive listing, the current inventory forms an extensive sequence of absolute dates with which the author has proposes a Pentapartite Division of Balearic Prehistory. This chronometric framework, which in the present edition shares equal status with the dating inventories, will be subsequently discussed in some detail; as well as graphically represented. For sequential purposes, the Pentapartite Division of Balearic Prehistory and its discussion are included after the examination of the various dating methods analyses studies. There is also a brief discription of each of the principal sites and stratigraphies forming the excavational contexts of the dating surveys.

From the beginning, due to the large number of analyses results involved in the inventory and the apparent rate at which the yearly number were foreseen probably to augment over the years, the original text adopted a system by which a permanent inventory could be maintained, and new Balearic dates added as in-coming information and data became available. Here in 1986, this recording system is still adequately functional in its structure, warranting its continued use for the addition of new data and other descriptive information.

As in the earlier inventory, a set of temporary calendar year conversion tables and curves (**Clark 1975**) based on Bristlecone Pine calibration of radiocarbon dates are provided the readers; should they be interested in the results of such conversion of radiocarbon dates into calendar years and, in turn, give some idea of the differences in converted dates as compared with conventional radiocarbon readings. Eventually, permanent conversion tables, the author believes, will come into universal use to cope more accurately with calendar year calibration needs.

In addition to the conventional radiocarbon technique documentation, three instead of two additional new chronometric methods of age determination are also briefly described and discussed: (1) the racemisation of aspartic acids in bone (**Bada and Schroeder 1973**), (2) the 14C carbonate dating of lime (**Stuiver and Valdren 1975**)(This latter survey is being further extended by a Belgium Series of 25 additional analyses results on carbonate dating of lime now in progress: **Van Strydonck in preparation**) and (3) the recent analyses results of uranium thorium dating of bone (**Ivanovich, Rae and Sweeting 1984**).

All three of these new methods have been applied extensively with interesting results, having been cross-correlated with conventional radiocarbon age determination from the **primary** Mallorcan sites of Muleta, Matge and Ferrandell-Oleza- the latter a newly excavated Copper Age open-air settlement- as well as other **secondary** sites included in the original inventory.

For overall continuity in the dating sequence and the analytical inventories and programmes available in the Balearic Islands generally, a list of spectrographic analyses of Balearic prehistoric metals is included (**Valdren 1979; Northover 1985**). The inclusion of a set of fossil pollen analyses results published earlier (**Valdren 1982**) is made to illustrate an important transitional stage in the Balearic chronological sequence.

The present edition is organised in the following manner. It is made up of five parts. **Part I: Methods of Analyses**; **Part II: Primary, Secondary and Tertiary Research Sites**; **Part III: Pentapartite Division of Balearic Prehistory; Part IV: Observations, Conclusions and Bibliography and Part V: Age Determination and Other Inventories**.

William H. Waldren, D.Phil. (Oxon)

Donald Baden-Powell Quaternary Research Centre,
Pitt Rivers Museum, University of Oxford,
Oxford OX2 6PN, England.

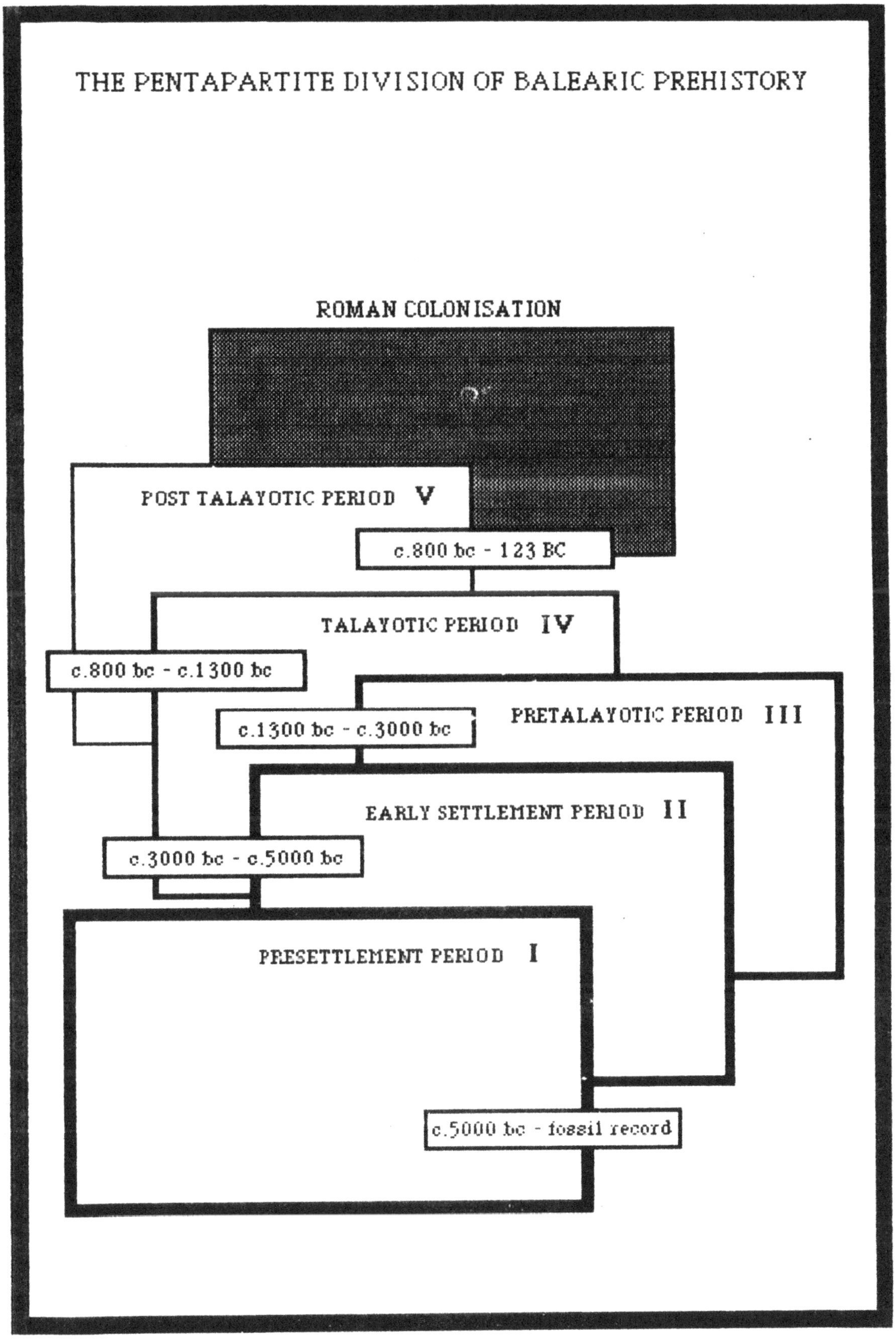
THE PENTAPARTITE DIVISION OF BALEARIC PREHISTORY
ROMAN COLONISATION
POST TALAYOTIC PERIOD V
c.800 bc - 123 BC
TALAYOTIC PERIOD IV
c.800 bc - c.1300 bc
PRETALAYOTIC PERIOD III
c.1300 bc - c.3000 bc
EARLY SETTLEMENT PERIOD II
c.3000 bc - c.5000 bc
PRESETTLEMENT PERIOD I
c.5000 bc - fossil record

FIGURE 2

Introduction

<<<<<<<<<>>>>>>>>>

Continental scholars and prehistorians, apart from a few exceptions, whether they have been particularly concerned either with early life forms (**palaeontologist/anthropologist**), or man and his cultures and environments (**anthropologist/archaeologist**), have shown a rather sporadic and mediocre interest or any significant gain to be earned from a long ranged study of Balearic prehistoric problems generally; much less attempting to associate or relate these problems with others farther afield. Probably the one reason for this, more than any other, has been this past poor assessment and under-estimation of the quality and the importance of scientific evidence and information to be gained from the islands as compared to the more extensively 'trafficed' climes of continental regions. In short, Balearic Island investigation as a source of important data and information has been considered up until very recently an ecological and cultural '**backwater**' from which only a narrow and limited perspective of prehistoric events in this corner of the Western Mediterranean could be expected or related in any significant way to events elsewhere. Another reason, no doubt, has been the lack of interaction and communication in the past on the part of those few workers, both, local and international -the author included- actually carrying out the research in the Baleares.

However with a more recent recognition of the importance of insular studies by a few continental prehistorians and scholars and the wotk done by them, mainly in the Eastern Mediterranean (**e.g. Evans 1973 and Sondaar 1973**), there has been a new awakening and stimulated general interest in the problems of insular prehistoric ecology and culture. At the same time, giving credit where credit is due, there have been a few exceptions in the past where investigators have looked to the Balearics as unusually rich areas for investigation (**e.g. Dorothea Bate 1909 and Margaret Murray 1938**). These were two such researchers whose insular work and scientific contributions in the first half of this century in the Balearic Islands have in more than one way inspired this current research and writing. As a potential matter of interest to the reader, a more detailed summary account of the history of archaeological research in the Balearics can be found in a subsequent section (**Part III**).

In recognition of the need to better define these lesser known Western Mediterranean insular geographic areas, a recent long-term research programme of excavation and dating archaeological and palaeontological sites and their contexts has been underway in the Spanish Balearics since 1962, using radiocarbon and other dating methods; the results of which are offered here.

Up until less than a decade ago, the inventory of absolute dates and resulting chronometric framework for Balearic prehistory emerging from the general research programme have centred mainly on the Island of Mallorca. However since 1981, the general survey programme in the Balearics has been extended farther afield to include dating results from the islands of Menorca and Formentera. For example, the small Island of Formentera recently has produced some interesting discoveries. There, a megalithic tomb structure has been excavated and some absolute dating information currently has become available (**Fernandez, Plantalamor Massanet and Topp 1976 and 1984**). Also several sites on Menorca, especially the Settlement Complex of Torralba d'en Salort (**Fernandez-Miranda and Valdren 1974**) have produced interesting dating series. And one cannot omit other local Mallorcan sites like the radiocarbon dating sequence from Son Fornes (**Gasull, Lull and Sanahuja 1984**) which also provides us with new and important data since 1981; significantly adding to the overall radiocarbon inventory.

The current inventory understandably, however, favours Mallorcan sites as it is here where the survey programme of absolute dating which forms the present chronometric framework for Balearic prehistory was initiated and most of the pertinent research carried out. As pointed mentioned earlier in the preface, the inventory as it appears in this edition is the most comprehensive published, and despite the fact that various interim listing have been available in published form for some time (**Valdren 1967, 1968, 1972 and 1981; Rossello Bordoy, Kopper and Valdren 1967; Rossello Bordoy and Valdren 1973; Fernandez-Miranda and Valdren 1974**), the body of the current inventory here in 1986 is much more extensive, having been augmented to include all available readings from the Balearics; including those published by other investigators.

The recording system, therefore, still retains its original function of forming the basis for a permanent catalogue of datings for the Balearic Islands. In essence, the system greatly simplifies the addition of new readings as information is received, and permanently fixes catalogue numbers for future use in quoting local dates.

Briefly outlined here, the system as used in this inventory consists of a series of abbreviations which designate (**a**) the island from which the dated sample originates, (**b**) the site name, (**c**) the laboratory conducting the analysis and the reading's official laboratory number, (**d**) the nature of the sample, (**e**) the local chronological period of which it is representative, (**f**) followed by the b.p. and b.c. dates (**uncalibrated**). Radiocarbon dates added later take on the inventory number that is closest to whatever chronological reading is already in the sequence. An additional lower case alphabetical letter is then fixed to the inventory number (e.g. **46, 46a, 46b** etc.). A more detailed description of the recording system is found in a subsequent section (**Part IV**) In this way, a continuing catalogue can be maintained as site data is received or as new stations are included.

Along with this present inventory, a set of tables and charts are provided for the interested reader, representing the optimal calibration curve for the conversion of conventional radiocarbon dates (**Clark 1975**) using 5568 year half-life. These simplified tables and charts are presented in place of the older Suess and Clark Bristlecone Pine calibrations (**Renfrew 1973; Renfrew and Clark 1974**) included in other earlier editions. Sooner or later, as pointed out recently (**Baillie and Pilcher 1983**), an universal conversion listing of all existing radiocarbon dates is necessary and no doubt eventually will become available. Until that time, the present inventory provides conversions in temporary supplementary table form along with the 1975 Clark Bristlecone Pine Conversion Charts for the reader's immediate conversion use .

The Pentapartite Division of Balearic Prehistory (**Figure 2**) with its various periods and their more universally associated contexts presented and examined in this paper are as follows.

Period I.	**The Presettlement Period**	**(Palaeolithic/Neolithic)**
Period II.	**The Early Settlement Period**	**(Neolithic)**
Period III.	**The Pretalayotic Period**	**(Neo-Chalcolithic)**
Period IV.	**The Talayotic Period**	**(Bronze Age)**
Period V.	**The Post Talayotic Period**	**(Iron Age)**

A general summary of each of the five periods, their subdivisions as seen in relationship to the available chronometric and other evidence are given here as an introduction to a more detailed examination and discussion in a subsequent section (**Part III**).

I

The Presettlement Period, the Pleistocene until 5000 b.c.

This period concerns the Balearic fossil record of extinct animal species, and briefly examines their ancestral types and that which is known about their age, origin, development and ecological environment prior to the coming of man.

It is the period that is of the longest duration and one which is delineated only by the evolution of the major, endemic Balearic mammal, ***Myotragus balearicus*** and smaller mammalian species ***Hypnomys morpheus***, a giant dormouse and ***Nesioties Hidalgoi***, a giant shrew.

The research and test materials illustrating this Presettlement Period come exclusively from the Cave of Muleta, apart from a brief consideration of the fossil record and some reference to particularly new findings; pertaining to these earlier species from selected other sites. It is this period the dating of which directly concerns and relates to two of the newer, less conventional dating methods subsequently discussed. Some brief reference is also made to the special method of hydraulic treatment used in the processing of the Muleta cave's sediments for faunal specimen extraction and sedimentological statistics (**Waldren 1982: Appendix 1B**).

II

The Early Settlement Period, circa 5000 b.c. to 3000 b.c.

This period is delineated by the earliest evidence of the arrival of man in the Balearic Islands as demonstrated in two of the primary research sites: the Cave of Muleta and the Rock Shelter of Son Matge. It also concerns the important aspect of man's coexistence with the endemic ruminant ***Myotragus balearicus***, circa 5000 b.c. to 3000 b.c. and beyond. The evidence of man and ***Myotragus*** in this period, however, is found without pottery present in the stratigraphies of the research sites. However, the author prefers to consider this period as **aceramic** rather than **preceramic**. Pottery was certainly in evidence elsewhere on the continent during this time, but its lack in the pertinent stratigraphy at Matge is due more, he believes, to the poverty of the particular levels of the site rather than the complete absence of a knowledge of local ceramic technology.

A more coherent account of the pertinent contextual, chronometric data and artefact evidence, representing these two periods from the primary research sites of Muleta and Matge is given elsewhere (**Part III**). There the geographic situation and the individual stratigraphies and the methodologies used in the excavations are examined more closely.

III

The Pretalayotic Period, circa 3000 b.c. to 1400 b.c.

<<<<<<<<<<<<<<<<<<<<<<<<<<<<<<<<<<<<<<<<<<<<>>

This period is represented mainly by evidence of human activities and artefact materials found in caves and rock shelters, but also encountered recently in open-air settlements, prior to local Bronze Age Talayotic megalithic building activities. The period has been up until recently considered synonymous mainly with the eneolithic period and the local culture of the caves and rock shelters. However as a direct result of current investigations, it can be demonstrated as also including a full-fledged Copper Age, involving well organised and architecturally rich open-air settlement. Burial customs are varied with multi-varied evidence supporting primary and secondary burials in caves-both natural and arteficial- as well as constructed tombs of different types.

In this case, the sites employed for the research are the three primary sites of Muleta, Matge, Ferrandell-Oleza Old Settlement and secondary ones of Muertos Gallard and Marroig with brief references to other selected locales where comparative chronometric data and materials have emerged. The period can be subdivided as follows:

A

Neolithic Early Ceramic Phase (NECP) circa 3000 b.c. to circa 2000 b.c.

<<<<<<<<<<<<<<<<<<<<<<<<<<<<<<<<<<<<<<<<<<<<<<<<<<<<<<<<<>>>

This chronological phase marks the earliest evidence of Balearic pottery technology and introduction of domesticated animal species of Sheep, goat, pig, cattle and dog, an some evidence to believe the presence of horse in the dated stratigraphies of Matge, Muertos Gallard and Ferrandell-Oleza.

In Matge, pottery is shown to be associated with the remains of ***Myotragus*** as well as domesticated animals, found in hearth and living zones of the site; thus demonstrating a combined set of associations, with special emphasis in the very late survival of ***Myotragus***, circa 2200 b.c. (**Burleigh and Clutton-Brock 1980**).

B

Early Beaker Phase (EBP) circa 2000 b.c. to 1600 b.c.

This chronological phase marks of evidence in the primary sites of the Rock Shelters of Son Matge and Muertos Gallard and the Ferrandell-Oleza Old Settlement of cultural contexts belonging to the **Bell Beaker Complex** associated with a workshop area, living zones on the one hand and open-air settlement on the other. Both are attested to by an abundance of Beaker wares of EBP type and other Beaker associated artefacts from the research sites similar to those found in continental sites, circa 2000 b.c. These various Beaker contexts from the primary, secondary and tertiary research sites are strongly supported by a coherent series of radiocarbon dates beginning circa 2000 b.c. to 1700 b.c.(**Table 9**).

While no <u>rich</u> copper ore resources have so far come to light, minor local sources are known to exist in the geology of the island of Mallorca, and it is during this phase that positive evidence has been found substantiating early metal working technology being carried out to some small degree. This is shown by the presence at Matge of what is believed to be a Beaker workshop which contained a bronze awl hoard, part of which can be seen to be in different stages of manufacturing, and Beaker decorated and undecorated, crucible sherds with heavy, interior copper oxide encrustrations (**Valdren 1979**). There is still other copper working evidence at the Ferrandell-Oleza Old Settlement (**Valdren 1984**).

The Beaker geometrically decorated pottery evidence representing this phase and the following one from primary, secondary and tertiary sites with 14C documentation- including a remarkable 14C series from the Ferrandell-Oleza Old Settlement (**Figure 7**)- and secondary sites with no 14C documentation, but with comparable materials are examined in detail in subsequent sections (**Parts II and II**). An introductory account of the chronometric dating results and analyses bronze samples from the research sites are presented there later with a more detailed presentations of the bronze analyses results (**Tables 19 & 20**), also offered in a subsequent section (**Part IV**).

It is at this point in the proposed chronological scheme that there appears to be need for some adjustment in the interface of this upper LBP phase of the Pretalayotic and the lower EBA phase of the Talayotic Bronze Age from that presented in the author's original chronometric framework in the earlier edition of this publication and the author's doctoral thesis (**Waldren 1981 and 1982**).

The major changes consist of what appears to be a prolongation of the Late Beaker Phase to circa 1300 b.c., accompanied by a later dated transition into the Early Bronze Age of the Talayotic Period, circa 1200 b.c. These adjustments are founded on recent radiocarbon datings and other supportive evidence from a number of archaeological contexts; sites like the Ferrandell-Oleza Old Settlement with late abandonment dates of the Old Settlement and the 14C dating of the intitial construction context of Talayot #1 (**T1**) of the Ferrandell-Oleza Younger Settlement; as well as a number of excavational findings elsewhere. These aspects are examined and discussed in more detail in appropriate sections (**Parts II, III and IV**).

C

Late Beaker Phase (LBP), circa 1600 b.c to 1300 b.c

This chronological phase is delineated by physical evidence and relevant levels at Matge, Ferrandell-Oleza and other sites based on comparative changes in pottery techniques in the form of the LBP wares, supported by radiocarbon dating. For example, stylistic changes in the methods of decoration and motifs of the Late Beaker ware fine wares as well as certain changes in the common pottery, which appear on the basis of composition later and cruder in character, along with what appears to be certain changes in the firing methods on the clay fabric of the pottery.

There is also quite a bit of evidence of increased metal working and circulation of the end products, as well as the introduction of a true tin bronze, having taken place at the end of the preceding phase and this one, circa 1700 b.c. to circa 1600 b.c.

As the Balearic Islands have no known tin resources and very little significant native copper sources, this aspect of metal working technology, its origin and sources of raw materials are some of the main aims of a recent Balearic series of quantitative spectrographic analyses of metal from the primary and other research sites currently in progress (**Northover 1986**). A preliminary quantitative report of the samples trace elements can be found in Part V

IV

The Talayotic Period, circa 1300 b.c. to 800 b.c.

The period comprises the Balearic Bronze Age and the time during which Talayotic megalithic building was first introduced and reached its apogee. It is during this period that most of the array of architectural building types including the Talayots, for which the mother culture is named, come into full use and highly social and economically stratified, well organised settlements were established.

There is a common use of tin bronze and radical changes in pottery style and technology. As stated earlier no significant copper resources are known- and especially none of tin- so, it must be assumed that commercial contact with Iberian Mainland was being conducted all through this period. This is evident in the many, particularly rich metal items of a given comparable typology that were buried with the dead during and toward the end of this period.

Another interesting factor during this period, while traditional Pretalayotic burial practices continue to be carried out, to some degree, in the earlier phase, still involving caves and rock shelters, other new burial customs involving predominantly cremation or token cremation communal burials in caves and rock shelters, but also in architectural depositories are introduced and carried out .

Here too, as mentioned in a preceding section, there are minor adjustments necessary in the early transitional delineations of this period, based on new findings.

A

Early Bronze Age (EBA), circa 1300 b.c. to 1100 b.c.

This chronological phase includes the earliest radiocarbon date available associated with a Talayotic building; although not actually dating the construction stage of the building itself but supposedly an occupation level (**Son Severa-Pula, SP37: Rossello Bordoy 1972**). Pottery of an established typology is one of the distinguishing features of this site and is comparable to a number of others. However, it is the new findings at the Ferrandell-Oleza Settlement Complex and Younger Settlement, circa

1200 b.c. to circa 1000 b.c.- as well as other supportive site contexts- which have charcoal radiocarbon dates and comparable artefacts to suggest a longer duration to the Pretalayotic Period as well as a later date for the Talayotic construction period. This new aspect is discussed in some detail in the appropriate text.

In the Matge rock shelter, the earliest recorded date for the site's abandonment as a Pretalayotic habitational area was circa 1400 b.c., while funerary pottery of the same typology associated with cremation fires of the Talayotic cemetery date closer to circa 1250 b.c. Completely new pottery styles appear in contexts there and in the settlements by this time, and there are some decided changes in burial customs entailing cremation as opposed to those of the Late Beaker Phase, where secondary burials without application of fire appear to predominate as demonstrated in the Cave of Son Marroig circa 1500 b.c..

The Ferrandell-Oleza Old Settlement with Beaker contexts circa 2000 b.c. are abandoned circa 1300 b.c. and the beginning of the EBA phase of the Talayotic Period. Therefore some adjustment in the interface of the Early Bronze Age phase also appear to be necessary during this stage, as well as an extension of the Pretalayotic chronometric boundaries; aspects considered in more detail subsequently.

B

Middle Bronze Age (MBA), circa 1100 b.c. to 900 b.c.

This chronological phase represents the apogee of the Talayotic cultural period in the settlements, with bronze in common use and the settlements as social entities at their most active with building activities at their most elaborate. The Matge rock shelter was completely abandoned as a living area and instead the East Enclosure of the rock shelter was converted into a cemetery; a function for which it continued to serve for some 1200 years, until well after the turn of the millennium. The Ferrandell-Oleza Old Settlement was also abandoned as a habitational area around 1300 b.c. and it appears that at least one of the Younger Settlements Talayots (**T1**), of which there are four surveyed so far, was just beginning as a construction during this phase, circa 1000 b.c.. Once again, correlations in pottery typology and radiocarbon dates inter-relate the two sites of Matge and Ferrandell-Oleza as well as those of other contemporary settlement sites.

C

Late Bronze Age (LBA), circa 900 b.c. to 800 b.c.

This chronological phase delineates a continuance from the former middle phase with extensive building activities continuing vigourously in the settlement sites in the early stages; including construction of defense walls, renovation of existing structures toward the end of the phase, circa 850 b.c., and exemplified by the construction of sophisticated structures like the Taula sanctuaries on Menorca. At Matge, pottery kilns are dated circa 860 b.c., confirming certain parallel pottery typological and compositional characteristics.

In some respects, it is within this phase that the first seeds of decline in the Talayotic mother culture can be seen; the Balearics at the end of this time coming into the accelerating sphere of influence of classical commercial enterprise and growing colonisation, as well as other influences from the Bronze Age and Early Iron Age cultures of Northern Europe; such as the Hallstatt.

V

Post Talayotic Period, circa 800 b.c. to Roman Colonization

The Post Talayotic Period is distinguished from the Talayotic Period by several major factors: (**a**) the general introduction of iron and its common use, along with other exotic material objects like glass and lead (**b**) distinct changes in burial customs along with a variation of these funerary practises, (**c**) radical changes in pottery styles and ceramic technology along with that of other artefacts, (**d**) and toward the middle and late phases (**MIA and LIA**) an easy access to classical imported wares and (**e**) a steady decline of the Talayotic building activities and tradition, particularly from the middle phase (**MIA**). It is also the period during which the Balearics are first colonised and Balearic mercenaries take an active part in both Punic and Roman military campaigns.

The primary and secondary sites used to chronometrically delineate and substantiate these aspects are Matge, Ferrandell-Oleza Younger Settlement and Muertos Gallard, where activities demonstrating this period are well represented. The period may be sub-divided into the following three phases:

A

Early Iron Age (EIA), circa 800 b.c. to 600 b.c.

This chronological phase is a transitional one during which time there is a radical change in burial customs, from cremation and earth burials to inhumation in quicklime toward its end; although there are signs of former traditional funerary customs having been carried on simultaneously.

At Matge an extensive 14C dating survey has been carried out to establish the duration of the local burial phenomenon of inhumation in quicklime and the oldest date here appears close to the end part of the phase, circa 690 b.c. (**Stuiver and Waldren 1975**), and showing its duration until after the turn of the millennium. A similar dating survey programme is also being carried out in the Ferrandell-Oleza Younger Settlement.

This phase also sees the common use of iron artefacts of Celtic typology and bearing strong resemblance to Continental European types, with which they indeed appear to be contemporary. There are also vestigial elements in the pottery typologies from the late phases of the Talayotic Period and toward the end of this EIA of the Post Talayotic the first clear signs of contact with the classical world. The adjacent island of Ibiza is colonised during this phase in 654 B.C. and trade with Mallorca is evident from the saturation of the local economy with classical pottery of all kinds. Glass beads first occur as well as some lead ornaments during this phase.

B

Middle Iron Age (MIA), circa 600 b.c to circa 400 b.c.

This chronological phase is defined by the increasing appearance and distribution of classical trade items throughout the local Talayotic settlements as well as other items similar to Continental European contemporary types, like the La Tene complex. The stratigraphy at Matge gives an admirable sequence of radiocarbon dates from this phase as does the Ferrandell-Oleza Younger Settlement and other research sites. It is during this time that building activities in the settlements appear to be limited to the renovation of the original Talayotic ones. Perhaps due to a decline in man power, for it is during this phase in particular that Balearic Punic trade increases and historical record chronicles substantial mercenary involvement in Punic campaigns throughout the Mediterranean; for example as recorded by Diodorus of the recruitment of some 5000 Balearic slingers made in 406 B.C. for Carthaginian campaigns against Sicily.

C

Late Iron Age (LIA), circa 400 b.c. to circa 123 B.C.

This chronological phase sees the continuance of the social and economic developments already referred to in decline in the former chronological phases of the settlement sites and other stations and is terminated by the Roman Colonisation by Cecilus Metelus. The phase more than any other demonstrates the overall marked decline in the social and economic structures within the indigenous settlements, although these continue several hundred years after the turn of the millennium.

I

METHODS OF ANALYSES

A number of different dating methods and other types of analysis have been used in the various research sites examined. Some of analytical techniques are new, and some are more conventional methods employed for chronological interpretation. In all events, each one has been selected either because of the special nature of the site's deposition, its materials or as occurs in the case of racemisation of aspartic acid analysis, the stability of the deposit's internal atmospheric conditions. On this occasion, the stability of the cave's thermal atmosphere was a special requisite that seemed to have been ideal in the Muleta site for the use of this particular method.

The various methods used in the inventory and examined below are the following:

1. **Conventional 14C dating of organic materials**
2. **14C carbonate dating of lime and quicklime burials**
3. **14C dating of cave calcite formation (cores)**
4. **Racemisation of aspartic acids in bone**
5. **Uranium thorium dating of bone**
6. **Paleomagnetic dating of travertine cave floors**
7. **Palynological analysis in the cave deposit levels**
8. **Stratigraphical and sedimentological analysis and other standard excavational procedures.**

For obvious reasons lengthly artefact comparisons and chronological correlations discussions related to all the different analytical methods and results used cannot be given here. Besides, these have been discussed in ample detail in the appropriate sections of the text of the author's published Oxford thesis ans well as in separate scientific reports, and only the most pertinent description can be included in this present publication. Nevertheless, it is necessary occasionally to outline for descriptive reasons or explanatory ones, briefly here, a few examples of the types of correlation that have been possible.

For example, there is a remarkable agreement between radiocarbon readings and artefact evidence from five widely dispersed research sites, all examined and discussed subsequently (**e.g. Ferrandell-Oleza Old Settlement, Muleta, Matge, Muertos- Gallard and Cotxera**).

These sites in their particular contexts are representative of a particular prehistoric period; one which embraces an especially polemic area in the Balearic prehistoric sequence; that of the local Bell Beaker problem found within the Pretalayotic Copper Age. The radiocarbon dating series available from these quite different and widely geographically separated sites, and the stratigraphies they contain, have produced almost identical pottery and other artefact assemblages. But what is of special interest in this particular example is the resulting chronometrical sequence with which these sites provide us. The materials used in dating the various sites and their contexts were either charcoal or burnt bone. These dates in the site order above were: 3950 b.p. (**2000 b.c.**) ± 65 years (**SFO-OS 45a**); 3910 b.p. (**1960 b.c.**) ± 120 years (**SM47**); 3820 b.p. (**1870 b.c.**) ± 120 years, (**ABSM46**); 3790 b.p. (**1840 b.c.**) ± 80 years (**AMG54**); 3790 b.p. (**1840 b.c.**) ± 90 years (**SFO-OS47b**); 3750 b.p.(**1800 b.c.**) ± 120 years (**CX44**) (also see **Table 9**).

At casual galance, the close correlation of these dates becomes evident, delineating the duration of some 200 years between sites for the Beaker contexts. As will see subsequently, this close correlation becomes more meaningful when we examine the individual series from each of these sites and find the same sort of similarity.

Another excellent example of good correlation between dates have taken place, (however this time using two different methods on one material: bone from the endemic, antelope, ***Myotragus balearicus***,), can be seen from the results obtained from RAA and conventional14C datings of a single bone from the cave of Muleta. The analytical processes in this case were conducted in different laboratories with the results of 16,850 b.p. (**14.900 b.c.**) ± 200 years (**SM66**) and 16,335 b.p. (**14.385 b.c.**) ± 415 years (**SM65**). In this case, considering depositional nature of the Muleta site, where one never could be sure in taking a sample if any single bone was from the same animal and also considering the high antiquity of the samples as well as the widely different techniques employed, this is a close correlation.

In both the examples cited above, it is considered that well correlated results were obtained; results which do much to recommend these largely different techniques as excellent tools for chronological interpretation. At the same time, they illustrate that whatever the shortcomings in the methods themselves- and these do exist at present in all dating techniques- these are outnumbered by the advantages to be gained.

No doubt ever more sophisticated techniques for dating will continue to be developed in the near future, as in the cases of the relatively new methods (**RAA and UTD**) reported on in this 1986 edition and will still further extend our chronological / chronometric knowledge of the area. Apart from actinf as tools for general understanding and interpretation of prehistoric events, these new techniques can also be used to reveal shortcomingsthat possibly exist within and between the various methods themselves. In this way, they can contribute to the correction of apparent discrepancies in a method; similar to the shortcomings and the resulting recalibration in radiocarbon readings, brought about by the introduction of dendrochronolgy and the comparative study of the Bristlecone pines.

During the interim between1976 and the present up-dating of this inventory, many more dating series havebeen compiled mading possible the correlations and comparisons of local materials. These too have been the result of using different techniques, different laboratory facilities and different materials (e.g. bone, charcoal and carbonised grain, etc.). This approach has been especially useful where it has been possible to use test materials from within single site stratigraphies and where this has been done results demonstrate the advantages to be gained; not only, from site **test method diversity**, but also contextual **blanket datings**.

Many such similarly close correlations between individual sites, their materials and laboratory results received have been quite common occurrences throughout the Balearic survey. And apart from bringing home the advantages to be gained, they also illustrate, I believe, the quality of results that can be expected from the application of long range dating surveys using as many <u>different</u> analytical techniques as possible. There is one other major factor, the outcome of these interdisciplinary on-going dating surveys, other than the delineation of particular chronological problems and various points of interpretation, this is the way these surveys tie in and compare to those from the adjacent mainland and other continental areas. This latter factor has been examined and discussed in a number of recent publications (**e.g. Fernandez-Miranda and Waldren 1974; Almagro et al. 1977; Deya Conference of Prehistory, editors, Waldren, Chapman, Lewthwaite and Kennard 1985 and the Oxford International Western Mediterranean Bell Beaker Conference, editors,Waldren and Kennard 1986)**. Here, the correlations are other than locally inter-site or intra-site in character, but actually suggest more far reaching parallels.

The Test Materials Used in the Current Analyses Surveys

<<<<<<<<<<<<<<<<<<<<<<<<<<<<<<<<<<<<<<<<<<<<<>>>

Along with the use of many analytical techniques, a wide variety of test materials have been used in the individual series represented in the inventory. In the majority of cases, it has been a matter of using whatever particular test materials were available from excavation. Nevertheless, wherever and whenever possible more than one material and method have been used to obtain the dating results.

The materials used for the various analyses are listed and briefly described in the following list:

1. **Domestic animal and human bone (burnt and unburnt).**
2. **Wood charcoal**
3. **Carbonised grain or similar organic material**
4. **Charcoal powder**
5. **Carbonate in organic materials (e.g. bone in quicklime)**
6. **Carbonates in calcite formations (travertine, stalagmites, stalactites, etc.**
7. **Animal coprolites (*Myotragus balearicus*)**
8. **Travertine and clays (palaeomagnetics)**

In the cases where the analysis of more than one material from a single stratigraphical horizon has been used, this has been done, as pointed out earlier, in order to compare one method's results with those of another (e.g. **Stuiver and Waldren 1974; Strydonck in preparation**). In other instances, a number of laboratory facilities (e.g **Yale University, the Smithsonian Institution, the University of Washington at Seattle, the British Museum, Harwell Atomic Energy Laboratories and University of Oxford**) were used for testing the materials so that any, perhaps, significant differences in the result or technique could be compared for possible misinterpretation of a context.. In still other examples, the particular laboratory or other facility used may have been the only one equipped to conduct a specific analysis.

This type of cross-checking and compiling of dating results is best illustrated in the programme being carried out at present in the **Muleta** cave in which ***Myotragus*** bone from cave deposit's various levels have been dated by conventional 14C method, while other samples from the same or different cave levels have been tested by other methods (**e.g. the Racemisation of Aspartic Acid in Bone, Bada 1975** and **Thorium Uranium Dating of Bone, Ivanovich, Rae and Sweeting 1984**).

The results of these separate series of analyses are included in the text of this inventory below. These are presented along with some description of the techniques themselves and with reference to pertinent publications as well as remarks on the test materials utilised in the surveys. Although two of these particular publications have not actually been co-authored, the present author has been either directly associated with the collection of the test samples or involved in some other capacity with the research team carrying out the Muleta and other site surveys.

In other instances, the test samples were collected by both the scientist conducting the analysis in company with the author. However in the majority of cases, the responsibility for the collection of the test samples has fallen on the author alone; using the collection methods recommended by the scientist carrying out the analysis. Regardless, all the samples have been collected with the utmost care as to the geographic and stratigraphic context, and the time gap between their collection and the analyses themselves has been as short as possible in order to ensure the minimum possibility of contamination, and thus obtain the best possible results.

Most of the inventory is comprehensive. Some of the conventional radiocarbon dates included, however, do originate from sites excavated by someone other than the author, though again a large percentage of these have been arranged through the his auspices in collaboration with the excavating scientists (**e.g. Rossello Bordoy and Waldren 1967 and Rossello Bordoy 1973**). The radiocarbon list also includes a number of results from independent datings with no collaboration and which have been published as data from independent excavations (**e.g. Rossello Bordoy and Frey 1964; Rossello Bordoy and Lilliu 1963; Tarradell and Woods 1967 and Gasull, Lull and Sanahuja 1984**).

Although these latter dates form the smaller percentage of the overall inventory, they give very important chronological and contextual data and do much to shed light on some of the problems, confirm or support findings and conclusions reached in the more extensively dated contexts at Muleta, Matge and Ferrandell-Oleza; as well as providing supporting evidence for the delineation and other interpretations included in the larger chronological framework created by the inventory itself.

In the inventory listing, the dates are presented in chronological order so that a single body has resulted- although illustrated here with large period divisions without any subdivision into phases as found in the text (**Part IV**)- to which future additions can be made without too much adjustment to the on-going listing. The inventory is also designed to aid the reading and general presentation of this publication as a whole. It does this by documenting the general remarks in the text and making possible a shorthand access to the chronology of the principal sites, as well as drawing together certain aspects of the author's own work and possibly that of others working in the area.

Individual inventory reference index cards are also included for more precise and pertinent information and data on each of the datings. These are designed more for future use than for the present, forming as they do computerised index cards locked within a more complex computer programme, where other contextual description index cards and other sample historical and quantitative data is stored. Because of the limited available space given to observations on the dating analyses index cards presented there, it is a matter of selecting the most pertinent data to appear by selecting this from the bulk of descriptive details attached to each analysis sample for inclusion at some future date. However, they are included in this current edition, despite the fact that some data and points of information in the observation sector are absent, in order to demonstrate briefly another facet of the recording system. Otherwise, the lack of more details in the observation sector is due to either a pertinency factor or because there is a complete lack of specific details from other recorded sites than those of the author.

Concerning the Types of Research Sites Used in the Inventory

The physical types of sites from which the test materials originate are nearly as varied as the materials and dating methods used. And while **blanket coverage** of contexts within author's own sites has been highly emphasised, there are some instances where only a small number of isolated or independent datings in his and other investigators' sites appear as the sole chronometric data available; thus resulting in a site having only one or possibly two radiocarbon dates. This has led to the reference in this inventory to a site **categorical** description in terms of either a **primary**, **secondary** or **tertiary** one. On the other hand , however, the physical types of sites themselves are equally varied in character.

The physical types of sites and site depositorial contexts used in this inventory and the resulting chronological framework can be divided into three major ones:

1. **Natural Caves containing animal and human remains or strictly traces of human activities.**

2. **Rock Shelters used by man and endemic and domesticated animals for shelter, other use or burial.**

3. **Open-Air Settlement Sites, where a large range of activities have been present.**

The geographic location of these different physical types of sites are widely dispersed geographically, mostly throughout the Island of Mallorca; although some reference is made to sites on the other islands of the Balearic Group (see **Figure 12 and the appropriate site sections found later in the text**). While widely dispersed geographically, these physical types also represent a good cross-section of the different possible stratigraphical conditions to be found in such locales. A closer examination of the individual research stations and the stratigraphical accounts used in this research are to be found in a subsequent section of this publication (**Part III: the Primary, Secondary and Tertiary Research Sites**).

The most reliable of these varied loci (in the sense of yielding the best stratigraphical conditions in vertical section; and hence chronological sequential succession) are the **caves** and **rock shelters**, as they normally contain the best preserved, least disturbed depositional contexts and other material sequences. **Open-air settlements**, on the other hand, are not as good for dating, mainly due to the wide range of agricultural and other activities, as well as the natural processes of soil erosion and redeposition etc., to which they have been exposed throughout the millennia. In short, Balearic open-air settlement situations are typical of most of those located in the Western Mediterranean Basin in general. This is to say that they are highly eroded by nature and therefore not at all famous for their vertical deposition; as open-air settlements are more apt to be in the climes of Northern Europe, where heavy top soil build up has taken place due to the growth and decay of heavy tree and other vegetation.

However in open-air site settlement situations, where the archaeological horizons and their contents have survived in either a vertical or horizontal stratigraphy locally, the dates obtained correlate remarkably well with those from the mainly vertical stratigraphical conditions found in caves and rock shelters.

Few mainland geographic areas as small as that of the Balearics have been the subject of as varied and extensive a chronometric campaign. All the same, the number of extensively dated individual, local sites is still relatively few. The results that have emerged, however, seem to form on the other hand a very coherent pattern of dates which can be relied on as a permanent source of reference; forming as they do a veritable **yardstick** for Balearic prehistory. They also do much to encourage confidence in the reliability of the techniques themselves, as well as the collection methods used; which in all but two cases show reasonably consistant statistical margins of error for the radiocarbon dates. In the two exceptions, aberrant results have been due to insufficient test materials being available at the time of the analyses; a shortcoming later rectified with the provision of larger amounts of test specimens; and so more acceptable and reliable readings.

The first dating materials in the Balearic survey of 1962 were human bones associated with those of the extinct ***Myotragus balearicus*** ; the islands' major prehistoric mammal, first described at the beginning of this century by Dorothea Bate (**1909**) and Charles Andrews (**1914**) as a rather peculiar, insularly developed and highly modified ***Nemorhaedine Antelope,*** endemic only to the Balearic Islands of Mallorca and Menorca. The animal's main evolutionary peculiarities and modifications are found in its skull, jaws and fore and hind limbs. Until the Muleta find and dating there, the animal was thought to have become extinct some 40,000 to 20,000 years ago and human occupation of the islands to have taken place no earlier than 2000 b.c. Further details concerning this animal as a chronological date setter and its relationship with man are found in a subsequent section (**Part II and III**).

DENDROCHRONOLOGY

Suess Bristlecone Pine Calibration and the Clark Optimal Calibration Curves for the Coversion of Conventional Radiocarbon Dates into Calendar Years in Regard to the Balearic Radiocarbon Inventory

The conversion of conventional radiocarbon dates into calendar years is still viewed with some reservation at the up-dating of this inventory in 1986. Since Suess first presented his calibration chart (**Suess 1970**), there have been several other modified 14C date conversion curves (**e.g. Ralph et al. 1973; Damon et al. 1972; Switzur 1973 and Clark 1975**), each of which present different variations of curve patterns for converting conventional 14C dates into BRCP calendar years. The author prefers to use R.M. Clark's (**1975**) conversion curve because it proports to modify most of the curves (**kinks and wiggles**) present in other conversion charts. It accomplishes this by establishing more data points for the plotting of the curve used in the determination and conversion of radiocarbon ages into calendar years; therefore eliminating many of the problems which arise of having to make delicate decisions as to the exact point where a 14C date, when projected, may strike one of the many "kinks and wiggles" found in other charts. Projection of 14C dates with older charts would often result in considerable differences in converted calendar year dates, when 14C determinations were closely grouped together; e.g. when they are separated by as little as ten years. As can be seen the dates listed in the present inventory frequently fall within this range of ten years differences in sample dates. For this reason mainly, the Clark (**1975**) conversion curve is much easier to use effectively; until a ratified curve, such as that suggested recently (**Baillie and Pilcher 1983**), is agreed on **officially**.

Undoubtedly considerable concern will occur when a general conversion does take place, as it will surely make for major alterations in existing chronological frameworks and particularly within the current Balearic divisions of prehistory. Nevertheless, this has to be accepted, and it seems more likely than not that it will undoubtedly have a very large effect on the author's own chronological framework presented here. As the late Gordon Childe among others recognised such chronological systems are necessary as working hypotheses for without them there would be naught but chaos. And whether or not they survive for very long is not really the important factor, but rather that a particular system division forms the best possible working hypothesis.

Concerning the Present Inventory System

Apart from providing us with the recalibrations essential for individual Balearic 14C datings into **BRCP** calendar years from circa 100 bc back to 4000 b.c., the calibrations also enable us locally to consider groups of dates within the various Balearic prehistoric periods over that or any given shorter time. This has been done locally by grouping two or more dates into a ***group unit*** (**groups 1-39**) as they appear in Table **1** (at intervals of 100 years for the most part) but extending longer in the last group unit (**47**). In the case of Table **2**, it shows groups of 14C dates in 14C years with their calendar year calibrations, and from it we can get some idea of the effect that such adjustments will have on the various dates of the Balearic prehistoric cultural periods individually. However, for the present no attempt has been made to make accurate evaluations of individual dates within their statistical errors, which should be calculated for each plus and minus probability factor. This should be done in the future with a separate study, considering the very large number of radiocarbon determinations involved in the Balearic inventory. Nonetheless, this general conversion of group dates, using Clark's tables and curve, is interesting as it gives a good notion of the

Table 1. Super Groups
Formed by Individual Groups of Analyses
Their Relationship to the Divisions of Balearic Prehistory
in
Terms of 14C Dates and BRCP Dates

Group No.	No. of Dates	14C-14C	BRCP Calendar Years
The Post Talayotic Period - Iron Age			
Super Group I ******************	0-26 55 total	**0-800 b.c.** 1950 A.D.-2750 b.p.	**0-975 B.C.** 1950 A.D.-2925 B.P.
The Talayotic Period - Bronze Age			
Super Group II *******************	27-37a 22 total	**800 b.c.-1400 b.c.** 2750 b.p.-3350 b.p.	**975 B.C.-1710 B.C.** 2925 B.P.-3660 B.P.
The Pretalayotic Period - Eneolithic/Neolithic			
Super Group III *******************	38-51a 22 total	**1400 b.c.-3000 b.c.** 3350 b.p.-5050 b.p.	**1710 B.C.-3885 B.C.** 3660 B.P.-5835 B.P.
The Early Settlement Period - Neolithic			
Super Group IV ******************	31-47 4 total	**3000 b.c.-4800 b.c.** 5050 b.p.-6750 b.p.	**3885 B.C.-4845 B.C.** *5535 B.P.-7500 B.P.

* Estimation as date in 14C b.c. is out of range of BRCP calibrated dating.

Each of the above Super Groups and the group units as well as individual dates of which they are made up are plotted on enlarged charts (Figures 4-7) to give a better idea of their relationship to the Clark (1975) chart (Figure 3). The first Suess calibrated conversion curve along with that of Switzur (1973) and Damon (1972) can be seen in still another chart (Figure 2) and are included here for comparative purposes.

TABLE

1

Table **2.** Grouped 14C Dates in 14C Years
Grouped Calibration in 100 Year Intervals
into
BRCP Calendar Years

Group		Total and Inventory of Analyses	14C-14C b.p.	b.c.	Calendar Years B.C.	B.P.
0-1	**8**	0-1g	1950-2050	0- 100		
1	**5**	2-3c	2050-2150	100- 200	95- 205	205-2155
2	**6**	4-10	2150-2250	200- 300	205- 400	2155-2400
3	**4**	11-12a	2250-2350	300- 400	400- 450	2400-2440
4	**7**	12b-15a	2350-2450	400- 500	450- 600	2440-2550
5	**14**	16-21	2450-2550	500- 600	600- 800	2550-2750
6	**5**	21a-23a	2550-2650	600- 700	800- 880	2750-2830
7	**6**	23b-26	2650-2750	700- 800	880- 975	2830-2925
8	**5**	27-29a	2750-2850	800- 900	975-1100	2925-3050
9	**4**	29b-31	2850-2950	900-1000	1100-1250	3050-3200
10	**6**	32-34c	2950-3050	1000-1100	1250-1385	3200-3335
11	**3**	34c-35a	3050-3150	1100-1200	1385-1495	3335-3445
12	**2**	36-36a	3150-3250	1200-1300	1495-1595	3445-3545
13	**2**	37-37a	3250-3350	1300-1400	1595-1710	3545-3660
14	**3**	38-39a	3350-3450	1400-1500	1710-1835	3660-3785
15	**3**	40-41a	3450-3550	1500-1600	1835-1975	3785-3925
16	**1**	42	3550-3650	1600-1700	1975-2095	3925-4045
17	**3**	43-43b	3650-3750	1700-1800	2095-2230	4045-4180
18	**4**	44-46	3750-3850	1800-1900	2230-2385	4180-4335
19	**1**	47	3850-3950	1900-2000	2385-2520	4335-4470
20	**4**	47a-50	3950-4050	2000-2100	2520-2670	4470-4610
21	**1**	50a	4050-4150	2100-2200	2570-2850	4610-4800
22			4150-4250	2200-2300	2850-2970	4800-4920
23			4250-4350	2300-2400	2970-3095	4920-5045
24			4350-4450	2400-2500	3095-3245	5045-5195
25			4450-4550	2500-2600	3245-3370	5195-5320
26			4560-4650	2600-2700	3370-3485	5320-5435
27	**1**	51	4650-4750	2700-2800	3485-3580	5435-5530
28	**1**	51a	4750-4850	2800-2900	3580-3685	5530-5635
29			4850-4950	2900-3000	3685-3785	5635-5735
30			4950-5050	3000-3100	3785-3885	5735-5825
31			5050-5150	3100-3200	3885-3990	5835-5940
32			5150-5250	3200-3300	3990-4095	5940-6045
33			5250-5350	3300-3400	4095-4250	6045-6200
34			5350-5450	3400-3500	4250-4375	6200-6325
35			5450-5550	3500-3600	4375-4450	6325-6400
36			5550-5650	3600-3700	4450-4520	6400-6470
37			5650-5750	3700-3800	4520-4590	6470-6540
38	**2**	52-53	5750-5850	3800-3900	4590-4680	6540-6630
39	**1**	54	5850-5950	3900-4000	4680-4845	6630-6795
47	**1**	55	6650-6750	4700-4800	* *	circa 7500
	103	**Total**				

* Out of dating range of BRCP (estimated date) 14c (5568 yrs. Half-Life)

TABLE

2

CLARK RADIOCARBON CALIBRATION CURVES

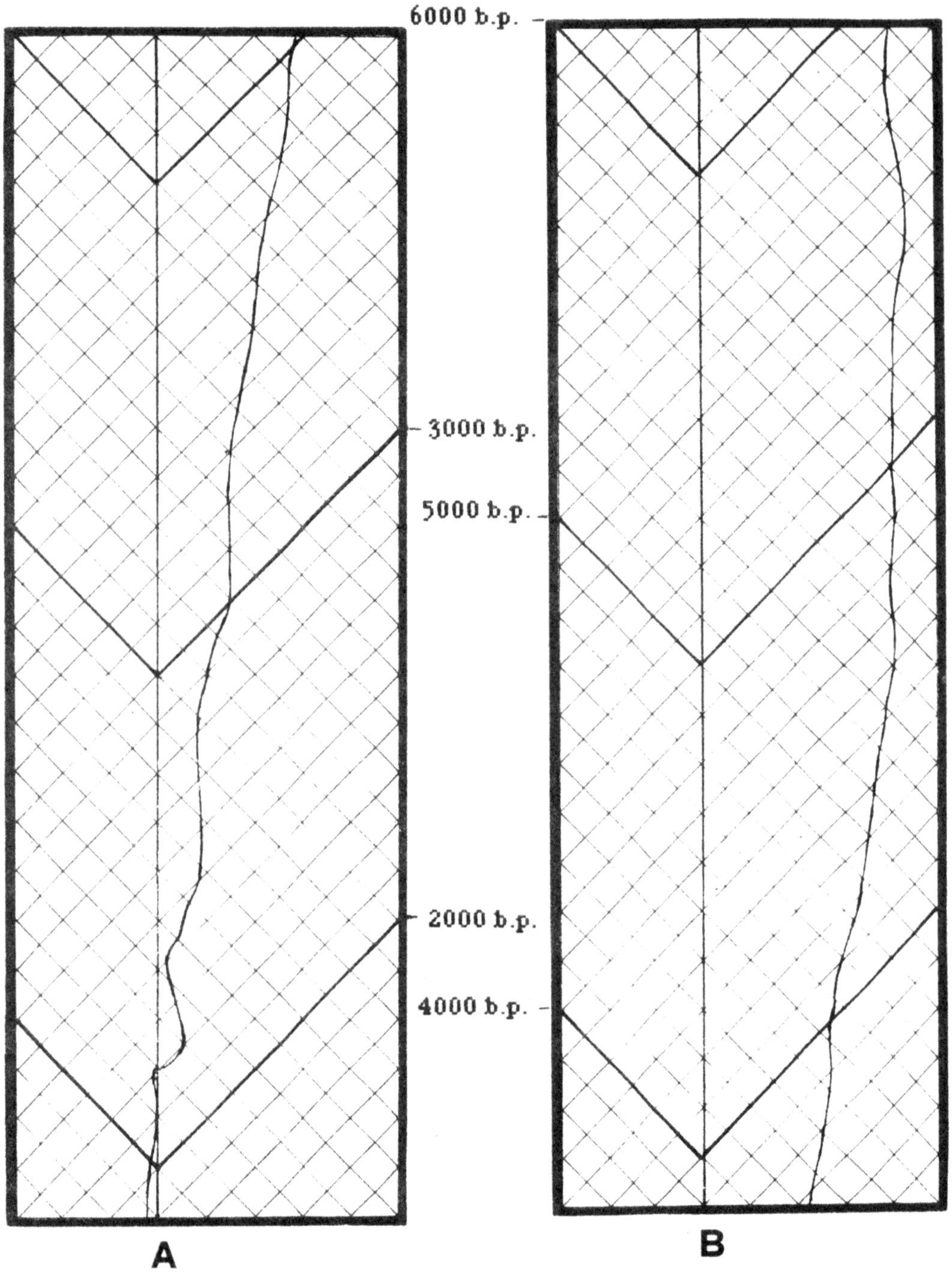

* Note: the Large 1000 Year Divisions are Subdivided into 100 Year Units.

FIGURE

3

magnitude of the alterations in the datings as they are related to the various Balearic periods; especially where closer comparison with Continental European calibrated dates may one day be necessary.

Concerning the Super Groups Used in this Publication

<<<<<<<<<<<<<<<<<<<<<<<<<<<<<<<<<<<<<<<<<<<<<<<<>>

The following table (**Table 1**) lists the various groups with serial numbers on the separate analyses used in them. These individual groups can be made into three distinct ***Super Groups*** as outlined in the table in the forms of ***boxes***: group units 0-7 form ***Super Group I***, group units 8-13 form ***Super Group II***, group units 14-30 form ***Super Group III*** and group unit 31 forms ***Super Group IV***. Each of the ***Super Groups*** corresponds with one of the Balearic Prehistoric Cultural Periods in the following manner:

1. ***Super Group I*** The Post Talayotic Period (Iron Age)
2. ***Super Group II*** The Talayotic Period (Bronze Age
3. ***Super Group III*** The Pretalayotic Period (Eneolithic-Neolithic)
4. ***Super Group IV*** The Early Settlement Period (Neolithic)

The ***Super Groups*** are important in the structure of the chronological framework used in the text of this publication. They also define the uncalibrated and calibrated limits of the **Balearic Iron Age, Bronze Age, Copper Age / Chalcolithic and Neolithic** phases.

In the second table (**Table 2**), the limiting dates in 14C years with the corresponding BRCP calendar years dates are given for each ***Super Group***, based on the various charts and tables. Further sub-divisions of the various periods into late, middle and early phases have been suggested in the text and their dates can be calculated within the known span of each of the ***Super Groups***, whether in calibrated or uncalibrated terms. Indeed some of the best support for the utilisation of the scheme of divisions and subdivisions used by the author is forthcoming from the radiocarbon evidence. In addition to the dating evidence, it greatly strengthens those comparative chronological proposals that might otherwise rely on typological considerations alone.

For the Balearic region, this has been the first serious attempt to bring into use chronometric evidence to arrive at more far reaching conclusions; rather than just simply an addition to the interpretation of individual sites; although this is still the most common objective to which such results are applied. On the other hand, while it is time indeed to use the existing dating evidence, amassed over the past 25 years, to attempt its correlation between the Balearic chronometric sequence framework and the same sort of sequences in other parts of Europe and the Mediterranean, it is not so simple a matter. We can expect that such a project as the present one, although very extensive and impressive locally on its own, will probably create new problems and in some respects retain the flavour of an isolated example of local interpretation and correlation. The reason for this among other things is partly due to the fact that many Mainland geographic areas still lack equally extensive dating surveys. Despite these shortcomings, I believe at least some first hypotheses can be put forward here; hypotheses based on the existing data and information as it does is found in those areas, even if these hypotheses are later prove in need of radical revision.

In any case, the system outlined here, despite the fact that no general dendrochronological recalibration of the radiocarbon dates has taken place, should not present any major problem in the overall radiocarbon dating correlations to result either locally or farther afield from such calibration. For example, at present, there appears to be <u>no</u> great discrepancies amongst the inter-site or for that matter intra-site correlations of the uncalibrated 14C readings. That is to say within the individual site stratigraphies themselves, or even in the chronological comparisons so far suggested farther afield in the areas of the Iberian Peninsula and rest of the Mediterranean Basin. On the contrary, the data and information are on the whole complimentary and supportive to such correlations wherever they occur. The main difficulty is one where there is more such evidence available from certain geographic areas than from others, or where there is little or no information at all. It is a certainty that many more critical geographic areas as well as important individual sites need **blanket** dating surveys before any real break through can be made in our interpetation and understanding of past events in many of these regions. In any case, with this new series, we are able to bring forward badly needed new chronometric data and information from a small geographical area, where none existed before, into the wider perspective of the Western Mediterranean as well as that of the rest of Europe in general.

RACEMISATION OF ASPARTIC ACID DATING OF BONE

Aspartic Acid Racemisation in Fossil Bones of *Myotragus balearicus* from the Muleta Deposit in Regard to Glacial and Post Glacial Temperature Differences

In 1971, as part of the author's continuing programme of chronometric dating, an additional survey, using the Muleta stratigraphy as a source of such data and other information, was suggested by Dr. John Ostrom of the Peabody Musuem of Natural History, Yale Unviersity, involving the then recently undertaken method of dating fossil bone, under development at the time by Dr. Jeffrey Bada of the Scripps Institute of Oceanography, University of California at La Jolla. The author, with Dr. Ostrom, arranged for the collection of a series of specimen samples from Muleta for analysis by the method.

In 1972, a research team consisting of Drs. J. Bada, S. Clark, J. Ostrom and K. Turekian and the author collected samples for the express purpose of radiocarbon dating and corresponding racemisation of aspartic acid analysles (**RAA**), the results of which are the subject of this section and which have been separately published (**Bada and Schroeder 1973**).

Summary of the Technique and Preliminary Results

The following section is a direct summary of the paper just quoted, and it is worth including here as the method is relatively new and there is still not much literature published on the subject. It is also included here because the method was applied to the Muleta cave stratigraphy and its materials and, therefore, a part of the chronometric data from the deposit. While the auhor is not a co-author of the quoted paper outlined here, he was and still is closely involved in the survey currently carried out in his research stations. Another important reason for its inclusion in this publication is to group together under one cover all or as much as possible the available age determination data concerning the research sites. At the same time, the RAA method is not too generally understood or for that matter accepted, although as will be seen, its use and correlation with 14C analyses using materials from Muleta have been remarkably successful; especially in regard to the site's lower stratigraphy, where the oldest of the cave's materials are located. Therefore, some descriptive discussion seems merited.

It is reasoned by the authors (**Bada and Schroeder**) that it is possible to determine the magnitude of temperature increases occurring in continental regions during the end of the last glaciation by establishing the degree of racemisation of aspartic acids in fossil bones of known age. The results that have been arrived at by Bada and Schroeder show an increase of 4ºC for the Mediterranean coast and 5ºC to 6ºC for East Africa. These estimates are believed to be reliable within 1ºC.

The basis of determining and estimating climatic changes that occurred on the earth during the glacial cycles is done in two ways: (1) by determining the $^{18}O/^{16}$ ratios in calcareous fossils and **(2)** the analysis of pollen and faunal distributions. Naturally, the interpretation of these are based on a number of factors, the absolute magnitude of temperature differences during the Pleistocene and interglacial cycles cannot be unevocally determined in the majority of cases. That which affects the $^{18}O/^{16}$ ratios found in calcareous fossils is the temperature fluctuations and isotopic changes in sea water, which results of the storage of greater amounts of '**isotopically light**' water in the continental ice sheets during extensive glaciation. The combination of factors affecting the production of the ratios of $^{18}O/^{16}$ are not fully understood. While pollen and faunal studies reflect the general kinds of vegetation, and in turn something of the general weather, these cannot offer accurate estimates of the prevailing temperatures, except that it was warmer or colder.

According to Bada and Schroeder in their paper estimations of palaeo-temperatures can be made by measuring the racemisation of amino acids. For example, only L-amino acids undergo slow racemisation, thus producing the non-protein acids, and depends on both time and temperature. So that if one of these variables is known, the reaction of the other should possibly be able to be used to determine the other. Therefore, hypothetically, the fossil materials that are found in good environmental situations -similar to those found in Muleta- can be used to estimate the age of a level and the fossil evidence by the extent of the racemisation of amino acids and palaeotemperatures evaluated by using other information. By dating the fossil evidence by C14 analysis, the racemisation reaction can be a method in which to estimate the average temperature that a sample has undergone since its deposition. For example Bada and Schroeder measured the racemisation amount of a 14C dated hippopotamus bone, and were able to show that the average of a warm spring (**Florisbad**) in South Africa was nearly constant for the past 40,000 years. They also were able to demonstrate that a 14C dated shell deposit from the Southeastern United States possibly had a severe temperature decrease in the average temperature during the last glaciation in this area of about 15ºC.

The authors' conclude that the most useful amino acid for making estimates of palaeotemperatuares is aspartic acid as it has one of the fastest racemisation rates of the stable amino acids. The half life of aspartic racemisation of bone is 15,000 to 20,000 years, if the ambient temperature is 20ºC. The reader should note that the temperature of Muleta cave has shown the average yearly temperature in the cave to be 19ºC to 19.5ºC, so that the cave deposit was initially suited for this research programme.

Table .

Analyses Results
14C Carbonate Dating Age
of
Post Talayotic Quicklime Burials in Mallorca

Lab No.	Material	Quadrant	Depth	13C (o/oo)	14C Ages
Post Talayotic Lime Burial Strata in Chronological Order					
QL-8	Ch.	M.M. 36	45 cm.	-23.0	2,080 yrs. b.p **130 b.c. ± 90 yrs.**
Y-1791	Ch.b	M.M. C	35 cm.		2,180 yrs. b.p. **230 b.c. ± 100 yrs.**
QL-9	Ca.	M.M. 36	30 cm	-16.8	2,200 yrs. b.p **250 b.c. ± 100 yrs.**
Y-1859	Ch.b	M.M. W	20 cm.		2,230 yrs. b.p. **280 yrs. ± 80 yrs.**
Y-2672	Ca.	M.M. 35	80 cm.	-13.6	2,240 yrs. b.p **290 b.c. ± 80 yrs.**
QL-22	Ca.	M.M. 44	20 cm.	-11.7	2,260 yrs. b.p. **310 b.c. ± 60 yrs.**
QL-5C	Ch.	M.M. 8	60 cm.		2,290 yrs. b.p. **340 b.c. ± 100 yrs.**
Y-2669	Ca.	M.M. 33	40 cm.	-18.3	2,400 yrs. b.p **450 b.c. ± 80 yrs.**
QL-10	Ch.	M.M. 33	90 cm.		2,480 yrs. b.p. **530 b.c. ± 70 yrs.**
Ql-6	Ch.	M.M. 33	75 cm.	-23.4	2,520 yrs. b.p. **570 b.c. ± 80 yrs.**
Ql-26	Ch.	M.M. 22	55 cm.		2,520 yrs. b.p. **570 b.c. ± 80 yrs.**
QL-4	Ch	M.M. 37	90 cm.	-22.5	2,540 yrs. b.p **590 b.c. ± 100 yrs.**
QL-27	Ch	M.M. 35	90 cm		2,640 yrs. b.p **690 b.c. ± 100 yrs.**
QL-11	Ch.	M.M. 35	100 cm.		2,700 yrs. b.p. **750 b.c. ± 80 yrs.**
QL-7	Ca.	M.M. 33	75 cm.	-19.9	2,730 yrs. b.p. **780 b.c. ± 100 yrs.**

TABLE

4

Pretalayotic

Fire Levels Forming Lime Carbonate in Chronological Order

QL-5	ca.k.h	M.M. 37	95 cm	-20.9	3,350 yrs. b.p. **1,400 b.c. ± 60 yrs.**
QL-5a	ca.k.h	M.M. 37	115 cm	-19.9	3,420 yrs. b.p. **1,470 b.c. ± 100 yrs.**
QL-24	ca.k.h.	M.M. 44	200 cm	-21.4	3,670 yrs. b.p. **1,720 b.c. ± 70 yrs.**
QL-5b	ca.k.h	M.M. 37	250 cm	-19.0	3,970 yrs. b.p. **2,020 b.c. ± 100 yrs.**
QL-23	ca.k.h.	M.M. 43	285 cm	-17.7	4,020 yrs. b.p. **2,070 b.c. ± 50 yrs.**

$$13C = \frac{(13C/12C\ \text{sample} - 13C/12C\ \text{standard})}{13C/12C\ \text{standard}} \times 1000$$

Legend of Abbreviations

QL = Quaternary Isotopic Laboratory
University of Washington at Seattle, Washington
Y = Yale University Radiocarbon Laboratory
New Haven, Connecticut
Ca. = carbonate
Ch. = charcoal
b. = bones
k.h. = kitchen hearth

TABLE

4a

Table shows the 14C ages of the fossil bone used by Bada and Schroeder- based on collegen fraction- D/L aspartic acid ratios and average racemisation rate constant (**Kasp**). The porous bones such as vertebrae have a greater surface and are more porous, thus more likely to contamination by ground waters etc. Indeed, D/L ratios were found by Bada and Schroeder to be generally lower in porous bone components.

The research of Bada and Schroeder indicates an averge temperature increase of 4ºC following the last glaciation's termination in the Western Mediterranean. Other research by Emiliani who used oxygen isotopic ratios of marine shells (food debris) found in caves along the Mediterranean coast estimated an increase of from 5ºC to 6ºC in the average yearly temperature. Still another investigator (**Duplessy 1974**) used isotopic composition of stalagmite found in a cave in southern France estimates a glacial and interglacial temperature change of 4ºC. The Muleta samples agree very well with both these estimates.

Further research combining aspartic acid age determination with others from the conventional radiocarbon method is being currently carried out in the Muleta cave and the Matge rock shelter. Results of this on-going programme will be added to the general inventory as they become available. Especially interesting should be sample dating from the deepest levels in Muleta proper as well as from a new chimney recently discovered adjacent to the lower system but part of the same rock outcrop. Here, articulated ***Myotragus balearicus*** skeletons have been found. When linked with the results of 14C datings presently in process at the British Museum radiocarbon laboratory, these RAA analyses of these new presettlement ***Myotragus*** levels should prove quite important.

In the research programme to be examined in the next section, a possible correlation may well emerge between the results of the two relatively new methods of Racemisation of Aspartic Acids and that of Uranium Thorium. Also, when linked with still other surveys underway dealing with the dating of sediments by thermoluminescence associated with the contexts mentioned above, further collaboration of these lower levels may be possible.

URANIUM THORIUM AGE DETERMINATION

Dating of *Myotragus balearicus* Bone

Introduction

The following paper was first presented by the authors' (**Rae, Ivanovich and Sweeting 1984**) at the **Deya Conference of Prehistory....Early Settlement in the Western Mediterranean Islands and Their Peripheral Areas**. Its inclusion here is for the sake of continuity to the inventory of absolute dating methods used and the results obtained from the Balearic research sites as well as the project to include all such analytical results under a single format.

The uranium series disequilibrium method of dating has been used successfully to date material relevant to palaeoenvironmental and archaeological studies (**Schwarcz et al. 1981**; **Schwarcz 1980; 1982**). This material at present includes speoleothem, spring travertine, corals and shells (e.g. **Atkinson et al. 1978; Ivanovich et al. 1983**). Bones and teeth have also been dated by this method (**Szabo et al. 1969; Szabo and Collins 1975**), but with limited success and much controversy.

The advantage of uranium series dating is that that it affords a dating interval from 1.5ka to 350ka, unlike radiocarbon, which has an ultimate limit of 40ka and reduced reliability after 25ka, due to high sensitivity to contamination. It is based on the presence of natural uranium isotopes, ^{238}U, ^{235}U and ^{234}U, in rocks and solution in groundwaters. From the ratios of the activities of the parent uranium isotopes to the daughters, ^{230}Th and ^{231}Pa, the age of a sample can be calculated, if it remains a close system with no uptake or loss of radiouclides after the parent uranium isotopes have been incorporated into the sample.

Fossil bone also contains uranium, often in much larger quantities than other materials, because of the great affinity of bone apatite for the uranyl ion. As bone is one of the most commonly found materials in excavation, a direct method of dating bone would be immensely useful. However, bone presents several problems. It is not a homogeneous, predictable material, but an organo-mineral structure, varying in size, density and mineral content from one sample to another. Fresh bone does not contain uranium but after burial, uranium is absorbed from the environment, in particular from groundwater. The rate of absorption, the lengths of time over which this takes place, and whether bone remains a closed system must be known before the dating method can be applied reliably.

This paper describes preliminary work examining the relationship between uranium and bone in the cave of Son Muleta, and also uranium in the matrices from which the bone samples derive. Comparison between the uranium series disequilibrium ages and those derived from radiocarbon measurement is also presented.

Site Location and Background Information

The Son Muleta site is unique in Western Europe in the large number of fossils of one species, ***Myotragus balearicus***, found during excavation of the two fossiliferous caves. Over 2000 individuals were found in the lower cave, a solutional joint feature forming a natural pit trap, from which no animal could escape. This has resulted in an unbroken succession of skeletons, dated by radiocarbon from 42ka to 6ka before the present. The Muleta caves are situated on a headland west of Soller, in the Sierra del Norte of Mallorca. They are found in the Muschelkalk limestone overlying Triassic red sandstones, 221 m. above sea level. Full details of the site are given by Waldren (**1982**).

Materials for analysis by neutron activation and alpha spectrometry were taken from the lower cave. Part of one sector, 'X', had been left *in situ*, so samples of matrix, and any remaining bone, were taken from each horizon. Samples were also taken from remaining horizons in sectors 'E' and 'Z' (see **Waldren 1982** for stratigraphy). Other bones were taken from the archives at the Deya Archaeological Museum. Solid limb bone, humerus, femur, tibia and radius, were taken wherever possible.

Analysis of Samples: Theory and Method

Neutron Activation Analysis

Neutron activation analysis is a method by which uranium content, specifically the presence of ^{235}U, can be measured (**Mapper et al. 1982**). The sample bones were thinly sliced with a diamond saw, set in resin and polished, giving sample areas of $1cm^2$. A piece of polyimide film was attached to each sample, and thirty two were exposed to the neutron flux of 2 x 10 neutrons/cm^2/second in the DIDO reactor at

Harwell for 30 minutes and 5 minutes. Nuclei created by the fission of the ^{235}U are scattered into the polyimide film creating tracks in the plastic, visible after etching in sodium hypochlorite solution. These indicate the position and quantity of the isotope in the sample. The first, longer exposure showed the pattern of uranium distribution across the bone. The second exposure produced a less dense pattern in which the tracks could be counted and concentrations calculated. The sample results were divided into sectors, in order of depth; mean and standard deviation were calculated for them (**Table 6**). The samples were then regrouped into bone types (**Table 5**) and a similar exercise performed. Chi-square tests were used to see if there is any statisically significant difference between the data sets.

Isotope Dilution/Alpha Spectrometry

The isotopes of uranium, thorium and protactinium used for uranium series disequilibrium dating all produce alpha particles as a result of spontaneous radioactive decay. The alpha particles are of different energies characteristic of each isotope involved, and can be counted to give a measure of the amount of each isotope in the sample. A fuller account of the theory and methodology is given elsewhere (**Ivanovich and Harmon 1982**). In practice, the samples of bone and bone matrix taken from Muleta were dissolved in nitric acid, after which uranium and thorium were separated and purified by use of an ion exchange resin (**Biorad AG 1-X8**). The samples were counted for 2 to 3 days and the data analysed using the UTAGE program (**Ivanovich 1982**).

Results

Neutron Activation Results

Variation Between Sectors:

The highest uranium content values found are for the samples from 'AB', 'O' and 'Z' sectors, all of them having approximately 4.9 ppm U. There are reasons for this. The higher uranium content in 'O' and 'Z' sectors is probably related to the position of the sectors in the cave. 'O' sector is directly under the chimney, which is the main source of dripwater in the cave. 'Z' sector is the continuation of the cave beneath 'O' sector, and is also a relatively damp place compared with other sectors. Therefore the total amount of water going through the bones will be higher than in other parts of the cave. The bones in sectors 'O' and 'Z' are therefore able to scavenge uranium more effectively from dripwaters. Higher uranium content in bone samples from sector 'AB' is less easy to explain, being further down the cave, at the top of the ramp leading to 'X' sector. The bones here may have been from a more lightly packed, porous matrix allowing a larger quantity of water to circulate. It is also possible that the bones were at one time in 'O' sector and have been moved to the present place by the action of animals or rolling down the talus slope from sector 'O'. The actual site of the bone in relation to the cave walls may also be important as there is likely to be more water available nearer the cave walls due to the condensation of water.

The Chi-square test was applied to the two groups of data: sectors 'AB', 'O' and 'Z' and sectors 'CD', 'EF' and 'X'. at the 0.05 confidence level, this showed that there is a statistical difference between the distribution of the two groups. The Chi-square test is not a particularly powerful statistical test, dealing as it does with nominal data, but it does confirm the general impression that there are statistically significant differences between the fossil populations within the cave. However, the high degree of variability should be noted for sectors 'AB', 'O' and 'Z'. The overall mean for the 3 sectors is 4.9 ppm U with a mean standard deviation of 2.38 ppm U. For the other group, the overall mean is 2.96 ppm U, and the associated standard deviation is 0.56 pm; a much tighter grouping than observed for the samples with the higher uranium content.

Variation between bone types

The comparisons were made when dealing with different bone types. One was a comparison of mature bones with those from immature animals. The other was a comparison of the four different bones shown: radius, tibia, humerus and femur.

The difference between the mean figures for the immature bones and the mean for mature bones is quite striking: (**0.85 ± 0.36**) ppm U (**immature**) and (**3.89 ± 1.57**) ppm U (**mature**). The low uranium content observed for the immature bones is probably due to lack of mineralisation in the bones of younger animals. Until sexual maturity, the mineral content of bone, mostly hydroxylapatite, is laid down on the calcifiable matrix in an unordered state (**Little 1973**). Remodeling and a secondary calcification later take place so that in an older animal there will tend to be a higher ratio of mineral content to organic content. Uranium is deposited within the mineral fractions, interstatistically or by ion exchange with calcium (**Neuman et al. 1949; Altschuler et al. 1958**).

Mature bones therefore provide a greater number of sites for uranium deposition than the juvenile bones where calcification is not complete.

There also appear to be differences between the types of bone (**Table 5**), although this is statistically not significant. The Chi-square test shows there to be no difference between the populations of the bone groups at all levels of confidence. Although there seems to be a small amount of variation due to density of bone, this is obviously not great enough to be of significance.

Isotope Dilution/Alpha Spectrometry Results

The results of this work are given in Table . This series of analyses is still in progress so the results reported here are only a part of the eventual full set. Although the major precept of the study is to examine the uranium budget of the cave and in particular the relationship between uranium and bone, the data derived from the bone analyses were also used to calculate bone ages.

Uranium Concentration in the Cave Sediments

Bones - The uranium concentration in the bones shows the widest range of any of the sediments; from 1.69 ± 0.03 ppm U or sector 'Z' at 650cm depth, to 25.80 ± 0.44 ppm U for sector 'X', level 13. In the other bones from 'X' sector, however, there is very little variation. For the four samples X7C, X7E, X7G and X11 the mean is 6.46 ppm U with a standard deviation of only 0.84 ppm U.

In 'Z' sector, there is an apparent stratigraphic reversal of uranium content. The bone sample from 'Z' 650cm, (**U/Th age 44.8 ± 1.6 ka bp**) contains only 1.69 ± 0.03 ppm U whereas bone from 'Z' 525cm, (**U/Th age 25.8 ± 1.3 ka bp**) contains 13.10 ± 0.19 ppm U. There are two possible interpretations of the phenomenon. The bone in 'Z' 650cm may be in open system with uranium migrating out of the bone leaving it depleted An alternative interpretation is that the bone was exposed to uranium in groundwater for only a short length of time, or during a relatively drier period so it was able to absorb only a small amount of uranium. It was then closed off from the uranium source, perhaps by covering with later layers of sediment and bone. Over a longer time scale, Seitz and Taylor (**1974**) have found a similar stratigraphic reversal of uranium content in the bone beds of the Olduvai Gorge, where the uranium content of Bed II is far higher than Bed I beneath or Bed III above.

Soluble matrices - The uranium concentration in all the soluble matrices are very similar, with an overall mean value of 1.63 ± 0.22 ppm U. This is to be expected as they are all derived from a mixture of weathered limestone from the cave walls, deposited calcite and small bone fragments. Not all the bone could be separated from the matrices; the extent of this high uranium content apatite dust probably accounts for the variation between the sample.

Insoluble matrices - Only three insoluble matrices have so far been analysed, however it appears that the uranium contents are considerably higher than the soluble matrices, and possibly more variable. This is related to the varied origins of the insoluble component of the matrix: wind-blown dust and translocated surface soil.

Uranium Series Disequilibrium Dates for Bones

The dates obtained for bones by uranium series disequilibria measurements are compared in Table with radiocarbon dates obtained for bones of the same or close stratigraphic localities. The results of this comparison are intriguing.

Immediately noticeable is the close correlation of some of the uranium series dates with the radiocarbon dates for bones of comparable sector levels. The bones from which the dates are derived have two common properties; they are from 'E' and 'Z' sectors, not 'X', and they are all older than 20 ka bp. The apparent agreement between the ages obtained for the bones is actually quite surprising considering the young age of the bones (compared with the range of the dating technique). The period of uptake of 14C is short and finite - during the life of the animal. The uptake of uranium is a post-mortem effect associated with fossilisation which may take place over thousands of years. The high levels of uranium in very recent bones, for example sheep and goat bones (**Sector '0': 0-50cm**) associated with human occupation of the cave, indicate that there is probably an early stage of rapid uranium uptake, followed by relative stability. However, the 'rapid uptake' stage may be several thousand years. We would therefore expect the ages obtained by uranium series measurement to be too young and require formulation of a corrective model. It is probably significant that the uranium series dates that compare best with the radiocarbon dates are from the older,

Table 5: **Neutron Activation Values for Uranium Content of Muleta Bones.**
Division into Bone Types.

Sector/Depth Deviation	U. Conc.	Mean and Standard
Radius		
AB 200-300	3.85	
AB 250-300	7.58	
CD 150-200 (Im)	0.50	
EF 100-150 (Im)	1.21	x = 5.25 (Mature)
EF 200-300	2.82	s = 2.55
E 300	3.53	x = 0.85 (Immature)
O 50-150	6.58	s = 0.36
O 200	6.36	
X 200-300 (Im)	0.84	
X 250	2.06	
Z 450	9.25	
Tibia		
AB 200-300	3.37	
O 175	1.79	x = 2.85
X 200	2.62	s = 0.70
X 150	3.22	
X 200-300	2.80	
Z 300	2.28	
Various 0-50	3.85	
Humerus		
CD 0-50	3.37	
CD 150	3.21	
F 100-150	2.43	
F 100-200	2.80	x = 3.51
E 300	4.64	s = 0.86
O 0-50	3.83	
X 200 (Im)	0.77	
Z 400	2.40	
Z 700	4.58	
Femur		
AB 150-200	2.82	
O 50-150	6.68	x = 4.33
O 150	4.15	s = 2.02
X 250	1.96	
Z 600	6.02	

TABLE

5

Table 6: **Neutron Activation Values for Uranium Content of Muleta Bones.**

Division into Sectors.

Sector/Depth	Bone Type	U Conc.	Mean & Standard Deviation
"AB"			
200-300	Radius	3.85	
200-300	Tibia	3.37	x = 4.93
250-300	Radius	7.58	s = 2.30
'CD'			
0-150	Humerus	3.37	
150	Humerus	3.21	x = 3.13
150-200	Femur	2.82	s = 0/28
150-200	Radius (Im)	0.50	
'EF'			
100-150	Radius (Im)	1.21	
100-150	Humerus	2.43	
100-200	Humerus	2.80	x = 3.24
200-300	Radius	2.82	s = 0.88
300	Humerus	4.64	
300	Radius	3.53	
'O'			
0-50	Humerus	3.83	
50-150	Femur	6.68	
50-150	Radius	6.58	x = 4.90
150	Femur	4.15	s = 1.97
175	Tibia	1.79	
200	Radius	6.36	
	Tibia	2.62	
'X'			
200	Humerus (Im)	0.77	
200-300	Tibia	2.80	x = 2.53
200-300	Radius (Im)	0.84	s = 0.53
250	Radius	2.05	
250	Femur	1.96	
'Z'			
300	Tibia	2.28	
400	Humerus	2.40	x = 4.91
450	Radius	9.25	s = 2.89
600	Femur	6.02	
700	Humerus	4.58	

(Im = immature bones/not included when calculating mean values)

TABLE 6

Table 7: Results of Alpha Spectrometry Analysis of Muleta Bones and Cave Sediments (soluble and insoluble fractions).

Sample and Harwell code-Huth no.	U. conc. (ppm)	^{234}U/^{238}U	^{230}Th/^{234}U	^{230}Th/^{232}Th	Bone Ages U/Th	Ka bp 14C
X 7B (50 cm)						
Bone (2968)	9.94 ±0.19	1.463 ±0.025	0.145 ±0.006	75.1	23.3 ±0.8	7.3 ±0.1
sol. (2926XA)	1.55 ±0.03	1.083 ±0.017	0.908 ±0.021	14.0	- -	- -
Matrices						
insol. (2926B)	8.51 ±0.21	0.812 ±0.024	1.257 ±0.045	5.3	-	-
X 7C (65 m)						
Bone						
(2934)	5.74 ±0.10	1.415 ±0.018	0.142 ±0.019	1000	16.5 ±2.3	7.3 ±0.1
sol. (2927A)	1.50 ±0.02	1.099 ±0.015	1.848 ±0.020	12.4	-	-
Matrices						
insol. (2927B)	7.88 ±0.17	0.749 ±0.018	1.568 ±0.076	4.3	-	-
X 7E (100 cm)						
Bone						
(2932)	5.61 ±0.10	1.332 ±0.021	0.167 ±0.009	10.5	19.7 ±1.1	8.8 ±0.4
sol. (2928)	1.57 ±0.03	1.089 ±0.107	0.890 ±0.020	11.1	-	-
Matrices						

TABLE 7

insol.	7.16 ±0.15	0.768 ±0.019	1.568 ±0.078	3.4		
X 7G (145 cm)						
Bone						
(2909)	8.17 ±0.12	1.327 ±0.011	0.141 ±0.010	23.6	16.4 ± 1.2	8.8 ±0.4
sol. (2681)	2.27 ±0.021	1.179 ±0.014	0.425 ±0.014	5.6		
Matrices						
insol.	-	-	-	-	-	-
X 11						
Bone						
(2933)	6.35 ±0.11	1.437 ±0.013	0.197 ±0.009	3.4 ±1.6	23.5 ±0.6	15.9
sol. (2903A)	1.60 ±0.03	1.078 ±0.020	0.890 ±0.018	10.4	-	-
Matrices						
insol. (2930B)	6.599 ±0.141	0.802 ±0.020	0.727 ±0.027	3.8	-	-
X 12						
Bone						
(2969)	10.87 ±0.21	1.392 ±0.026	0.282 ±0.010	61.1	35.4 ±1.4	-
sol. (2925)	1.42 ±0.03	1.099 ±0.020	0.968 ±0.023	15.7	-	-
Matrices						
insol.	-	-	-	-	-	-
X 13						
Bone						
(2931)	25.8 ±0.44	1.3558 ±0.100	0.123 ±0.013	100	14.2 ±1.1	16.9 ±0.2

TABLE

7a

sol. (2935)	1.37 ±0.03	1.082 ±0.021	0.756 ±0.024	16.5	-	-
insol.	-	-	-	-	-	-
X 15						
Bone (2971)	18.24 ±0.34	1.451 ±0.025	0.213 ±0.008	86.	25.7 ±1.1	18.8 ±0.5
sol. (2678A)	1.61 ±0.03	1.094 ±0.022	0.847 ±0.020	15.0		
Matrices						
insol. (2678B)	6.04 ±0.13	0.726 ±0.021	1.574 ±0.092	5.4	-	-
E 355 (cm) ------------------------------ E 400(cm)						
Bone (2192)	7.91 ±0.12	1.184 ±0.011	0.193 ±0.020	19.0	23.2 ±2.3	23.8 ±0.5
sol. (2682A)	1.39 ±0.03	1.121 ±0.025	0.401 ±0.016	11.3	-	-
Matrices						
insol. (2682B)	3.65 ±0.08	0.857 ±0.020	0.532 ±0.013	11.1		
Z 525 (cm) ------------------------------ Z 475 (cm)						
Bone (2910)	13.10 ±0.18	1.168 ±0.011	0.213 ±0.009	44.4	25.8 ±1.3	26.0 ±0.6
sol. (2680)	2.15 ±0.04	1.139 ±0.021	0.342 ±0.013	2.7	-	-
Matrices insol.	-	-	-	-	-	-
Z 650 (cm) ------------------------------ Z 600 (cm)						
Bone	1.69	1.362	0.344	19.2	44.8	42.0

TABLE 7b

(2911)	±0.03	±0.032	±0.010		±1.6	±1.0
sol.	1.72	1.167	0.256	12.0	-	-
(2679A)	±0.03	±.019	±0.012			
Matrices						
sol.	6.47	0.769	2.909	4.8	-	-
(2679B)	±0.15	±0.020	±0.122			

N.B. - 14C dates are from nearest dated horizons.

TABLE

7c

lower horizons. A comparative study of the two dating techniques on the same bones is essential before further speculations are made.

The uranium series dates obtained for bones in 'X' sector correspond rather less well with radiocarbon dates for bones of the same stratigraphic levels. In four cases the uranium series dates are older than the radiocarbon dates, by 8 to 11 ka, in one case (X 13), 3 ka younger. The latter case seems to correspond with the post-mortem uranium uptake model described above. The others however, seem to show either movement of uranium out of the bones or possibly represent a real set of dates from older bones mixed with young material during sedimentation. A direct comparison of dating techniques on the same bone is again needed to verify this. The ***Myotragus*** bones are individually too small for conventional radiocarbon analysis, but it is planned to use the microcounting system of the Low Level Measurements Laboratory at AERE Harwell to overcome this problem.

Discussion and Conclusions

As the analytical work described in the paper is till in progress, any conclusions made must be tentative. However, several points seem apparent from the data reported here:

1. The position of the bone to be measured must be known in detail to relate it to the hydrological system of the cave. This is obviously a major factor in the amount of uranium a bone is able to scavenge and therefore the final concentration of uranium within the bone.

2. The condition and maturity of the bone and therefore the likely apatite content are of significance with respect to uranium uptake.

3. The size and density of the bone are relevant to the rate of uptake and the total amount of uranium any bone can absorb.

4. The amount of uranium with a bone is not directly related to the age of a bone, as shown in sector Z.

5. The uranium concentration in the bones and the matrices do not appear to be related.

6. Some of the uranium series dates on bone seem to be in good agreement with radiocarbon dates, without the necessity for correction for uranium uptake, which was initially expected. However, uranium series and radiocarbon dates must be derived from the same bone before they can be realistically compared.

Muleta is a perfect site in so far as it contains a large number of well documented and radiocarbon dated bones for comparison with the uranium series disequilibrium method. The results obtained so far are very encouraging. It is not possible to say at present whether a uranium series disequilibrium dating technique for bones of universal application will be developed, but we feel that the work described in the present paper may contribute to the knowledge of uranium systems necessary for the development of such a technique.

14C CARBONATE DATING OF LIME

Age of the Post Talayotic Quicklime Burials in Mallorca:

Introduction

This investigation was supported by the United States National Science Foundation (**Stuiver and Waldren 1975**). The initial sampling began as early as 1965, but has been intensified since 1969 and is still in progress. Here in 1986, as pointed out earlier in the text, a series of 25 analyses of samples from the Matge deposit and elsewhere are currently in progress; the results of which will in due course appear in a future edition of this on-going research.

The principal sites used have been the rock shelters of **Son Matge** and **Muertos Gallard** because of the character of their reservoirs. The frequency of such reservoirs throughout Mallorca and Menorca has probably resulted in their being taken for granted by most local investigators, as well as their having been plundered by numerous nonprofessionals. At the same time, there has been a general opinion that quicklime burials were a short-lived phenomenon. The contrary is now known to exist.

The purpose of the current research has been to establish (**a**) the time duration of the custom, (**b**) its role in the Balearic chronological framework and (**c**) possible origin of the practice and the influences involved (**d**) as well as details of the method of manufacture itself. (**e**) A further aim has been to date a material resource which is both abundant and frequent found and has been hitherto neglected as such. The study method for dating todate has consisted of comparing analyses results with those of conventional 14C charcoal dating.

The analyses results for the moment indicate that the quicklime burial rite originated at least by the Sixth Century BC, and perhaps even earlier, according to recent indications (**circa 680 b.c.**). There is also conclusive evidence that the lime burials continued well after the Roman Colonisation of Mallorca (**circa 80 b.c.**). This would suggest a considerable duration; one of 600 years or more in which the substance was employed for inhumation burials.

The Custom of Quicklime Burials, the Balearic Phenomenon and Its Possible Origin

Inhumations in quicklime, their site localities and stratigraphies, especially those of Matge and Muertos Gallard, have been discussed in detail in the text of the author's doctoral thesis (**Valdren 1982**). It cannot be too heavily stressed, however, that the origin of the custom remains, as far as is known today, a strictly Balearic prehistoric phenomenon.

As pointed out, its use as a burial custom or rite occurred during the period when historical Mediterranean navigation and trade were in full progress. In the last centuries of the Talayotic Culture, as it is associated with the megalithic architecture from which it gets its name, Phoenician, Greek and Roman influences become increasingly evident in the form of colonisation and trade. At the same time, other indications of contact and even trade with western and northern contemporary cultures also are in evidence.

The use of quicklime as a burial medium in place of earlier methods as discussed in Chapter V of the author's thesis is one of the remarkable events which delineates the largest part of the local Post Talayotic Period (**Iron Ages**). Nearly nothing is known as to the origin of the idea, but it can be assumed that exposure to the processes of quicklime manufacture occurred; perhaps as the result of contact with the Classical World.

At the beginning of the Early Iron Age, circa 800 bc, the Phoenicians in the Canaanute east had developed a new method of lining water storage cisterns with impermeable lime plaster for long term storage of rainwater. Only later true cement was developed and many water storage cisterns with such linings exist from the later years of the Punic city of Carthage.

It is a matter of historical record that the Balearic island of Ibiza was settled by the Carthaginians in 654-653 B.C. (**circa 700 bc**) The Balearics were also involved in wars between Carthage and Rome between the Fifth and Second Centuries B.C., an involvement which culminated in the conquest of Mallorca by Caecilus Metelus in 123 B.C..

There is however no evidence that the Phoenicians used quicklime for burials. It can only be postulated (**a**) that the technical knowledge may have been brought back by travellers or mercenary soldiers from an area where lime manufacture was being carried out, and (**b**) that its practical application in disposing of the dead was therefore localised. It also seems quite evident that the method must have taken on some **religious** as well as **practical** burial significance. The evidence and logic behind this hypotheis rests in the following: (**1**) there is a frequency of quicklime cemeteries throughout the Balearic islands of Mallorca and Menorca (none are as yet known on the Punic Colony of Ibiza on the Island of Ibiza) which shows a widely accepted use of the method. (**2**) The manufacture of the lime itself in large quantities for burial use demonstrates very definitely a form of ceremony in its preparation, along with the other necessary preparatory actions in the collection of fuel for the fires to make the lime. (**3**) There is a complete lack of any evidence of its having been used for constructional purposes. Hence, the custom remains very much a Balearic phenomenon, so far unknown elsewhere in the Prehistoric World.

The Research Method, the Results and General Information

Examination has been made of lime carbonates and incorporated charcoal to determine the feasibility of lime dating for this local environment. The majority of the collected samples in 1969-73 came from the **Son Matge** burial site in Mallorca.

Four charcoal samples from the Late Talayotic internment epoch (**Post Talayotic**) taken from directly underneath the lime burial strata were collected from several excavated sections and associated with iron artefacts. The laboratory numbers (**QL=Quaternary Isotopic Laboratory**) and the dates (**years b.p. and years b.c.**) of these samples are **QL-10**, 2,480 b.p. (**530 b.c**) ± 70 years; **QL-11**, 2,700 b.p. (**750 b.c.**) ± 80 years and **QL-26**, 2,520 b.p. (**570 b.c.**) ± 80 years, **QL-27**, 2,640 b.p. (**690 b.c.**) ± 50 years, respectively.

This places the transition from regular grave techniques (in this case cremation by fire burials) to lime inhumation approximately 2,500 b.p. The 14C age is equivalent in calendar years to 510-660 B.C., according to Masca calibrated tables.

Table **The Sites Sampled, their Ages and Other Information**

D/L Aspartic Acid Ratios (D/LAA) and Average Racemisation Rate (KASP)

Location	Sample No.	D/L	RAA Date	KASP
Muleta Cave Mallorca, Spain	UCLA-1704c	0.222	8,570 yrs. b.p. **6,620 ± 350 b.c.**	1.72
	UCLA-1704d	0.273	16,850 yrs. b.p. **14,900 ± 200 b.c.**	1.25
	UCLA-1704e	0.293	18,980 yrs. b.p. **17,030 ± 200 b.c.**	1.22
Lukenya Hill Kenya	UCLA-1709c	0.154	2,120 yrs. b.p. **170 ± 60 b.c.**	4.02
	UCLA-1709b	0.500	17,700 yrs. b.p. **15,750 ± 760 b.c.**	2.71
Olduvai Gorge Tanzania	UCLA-1695	0.316	17,500 yrs. b.p. **15,550 ± 1000 b.c.**	1.48

*** **Notes** The authors' (Bada and Schroeder) give no stratigraphica information for the sites. However, stratigraphical details of the Muleta samples can be found in the inventory presented below.

The underlined sample reading in Table 3 signifies an average value reading obtained from the analysis of two different pieces of bone.

TABLE

8

Table .

CARBON 14 RESULTS FOR THE BALEARIC PRETALAYOTIC PERIOD AS RELATED TO THE BELL BEAKER CULTURE

SITE/AREA	CONTEXT	bc 14C DATE	bp DATE	LAB.
THE EARLY SETTLEMENT PERIOD				
1. ABSM	ESP	4730 ± 120 bc	6680 yrs.	QL-29
2. SM	ESP	3985 ± 109 bc	5935 yrs.	KBN-640c
3. ABSM	ESP	3870 ± 360 bc	5820 yrs.	CSIC-176
4. ABSM	ESP	3800 ± 115 bc	5750 yrs.	I-5516
THE PRETALAYOTIC PERIOD				
THE NEOLITHIC EARLY CERAMIC PHASE				
5. AMG	ESP	2810 ± 50 bc	4760 yrs.	BM-1994
6. ABSM	ESP	2700 ± 120 bc	4650 yrs.	QL-988
7. ABSM	ESP	2143 ± 398 bc	4093 yrs.	BM-1408
THE EARLY BEAKER PHASE				
8. ABSM	EBP	2070 ± 50 bc	4020 yrs.	QL-23
9. ABSM	EBP	2030 ± 120 bc	3980 yrs.	CSIC-178
10. ABSM	EBP	2020 ± 100 bc	3970 yrs.	QL-5b
11. SFO-OS	EBP	2000 ± 65 bc	3950 yrs.	BM-1843
12. SM	EBP	1960 ± 120 bc	3910 yrs.	Y-2389
13. ABSM	EBP	1870 ± 120 bc	3820 yrs.	Y-2389
14. SFO-OS	EBP	1840 ± 90 bc	3790 yrs.	QL-1636
15. AMG	EBP	1840 ± 80 bc	3790 yrs.	Y-1789
16. CX	EBP	1800 ± 120 bc	3750 yrs.	I-5515
17. SFO-OS	EBP	1770 ± 35 bc	3720 yrs.	BM-1981
18. SFO-OS	EBP	1750 ± 30 bc	3700 yrs.	QL-1592
19. ABSM	EBP	1720 ± 100 bc	3670 yrs.	QL-24
THE LATE BEAKER PHASE				
20. ABSM	LBP	1620 ± 80 bc	3620 yrs.	CSIC-179
21. SFO OS	LBP	1540 ± 80 bc	3490 yrs.	QL-1859
22. ABSM	LBP	1530 ± 80 bc	3480 yrs.	CSIC-180
23. SMRG	LBP	1520 ± 80 bc	3470 yrs.	Y-1856
24. ABSM	LBP	1470 ± 80 bc	3420 yrs.	QL-5a
25. ABSM	LBP	1430 ± 50 bc	3380 yrs.	BM-1995
26. ABSM	LBP	1400 ± 60 bc	3350 yrs.	QL-5
27. CNC	LBP	1320 ± 80 bc	3270 yrs.	BM-1667
28. SP	LBP	1310 ± 60 bc	3260 yrs.	UP-1438
29. SFO-OS	LBP-EBA	1260 ± 80 bc	3210 yrs.	BM-2312
30. ABSM	LBP-EBA	1250 ± 100 bc	3200 yrs.	Y-2667
31. SFO-OS	LBP-EBA	1140 ± 70 bc	3090 yrs.	BM-1698
32. SILL	LBP-EBA	1130 ± 75 bc	3080 yrs.	HU-1716
33. SFO-OS	LBP-EBA	1120 ± 50 bc	3070 yrs.	HAR-3490

TABLE

9

The lime strata in the Matge burial grounds have a thickness of up to one meter in the Central Enclosure and up to 150 centimeters. in some areas. A multitude of bone remnants is left for 14C dating because of the high temperatures achieved during lime slaking, literally having baked them to a near porceline-like hardness, although cracking and warping them.

Isolated charcoal samples were found in two sections at a depth of 75 centimeters and 90 centimeters. in the lime layer, close to the contact zone of the Late Talayotic phases.

The ages of these samples: **QL-6**, 2,520 b.p. (**570 b.c.**) ± 110 years confirm that the transition of lime burials took place about 2,500 years ago.

Charcoal from shallower sections (**45 cms.**) gave an age of 2,080 b.p. (**130 b.c.**) ± 90 years (**QL-8**). To this can be added two bone charcoal dates from lime related burials at Muertos Gallard with ages of (**Y-1791**), 2,180 b.p. (**230 b.c.**) ± 100 years; (**Y-1859**), 2,230 b.p. (**280 b.c.**) ± 80 years respectively. In these cases, **Y=Yale University Radiocarbon Laboratory** and the samples dated before the laboratory was transferred to the Quaternary Isotopic Laboratory, University of Washington at Seattle.

Evidently, lime burials lasted generally until about 2,100 years before the present epoch which is equivalent to a calendar age of 80-110 B.C. In other areas, it was possible to date the custom to around 50 A.D. with Roman pottery of known chronology. This Post Talayotic episode could not have lasted much longer, as the Roman artefacts occur directly over the more dense lime levels. The curious phenomenon of lime inhumation was probably practised for five or six centuries at least <u>before</u> the Roman Colonisation in 123 B.C. forward.

Two pairs of lime carbonate-charcoal samples (**QL-8/9 and QL-6/7**) were available for a test of the reliability of 14C dates from carbonate. The ages of these pairs agree with each other within the limitations of statistical error (**Table 4**). The ages of three other carbonate samples (**Y-2669, Y-2672 and QL-22**) also fell in the period of the lime internment between 2,500 and 2,100 years b.p. (**550 b.c. and 110 b.c.**).

Two of the carbonate samples which give older dates, though found close to the contact zones, originate from older habitational levels of Pretalayotic origin. This older carbonate was formed when the limestone floor of the cave was decomposed by Pretalayotic kitchen fires. The sample **QL-24** is a clear sample of such '**habitational**' lime carbonate. Its age is 3,670 b.p. (**1,720 b.c.**) ± 70 years which is in excellent agreement with the age of charcoal from associated fires and pottery evidence of the same chronology. The other two older carbonate samples are also from associated charcoal and kitchen fires of Pretalayotic origin and are not part of the lime burial deposit. **QL-23**, 4,020 b.p. (**2,070 b.c.**) ± 50 years; **QL-5**, 3,970 b.p. (**2,020 b.c.**) ± 100 years.

These latter samples of accidental manufacture of lime by the firing of kitchen hearths have proven to be in accord with charcoal analyses of layers above and below the above samples, **Y-2682**, 3,820 b.p. (**1,870 b.c.**) ± 392 years and **QL-988**, 4,650 b.p. (**2,700 b.c.**) ± 120 years.

The 13C values listed in Table represent the deviation o/oo of the sample 13C/12C ratio from the PDB standard. The values 13C of the carbonate sample or **normal** 14C age average - 16.1 o/oo. The lime at Son Matge has probably been made from limestone of the shelter wall which has a value - 13C of -1.4 o/ooo. If the older samples or carbonates already mentioned have been caused by incomplete degassing of the limestone during lime fabrication, the expected 13C values would be higher.

Age differences of 900 to 1500 years could be caused by the addition of up to 20% bedrock limestone, resulting in an increase maximum of the 13C ratio of 3 o/oo. But the 13C values of the older lime samples average -19.4 o/oo; 3.3 o/oo below the -16.1 o/oo found for the carbonate samples with the proper 14C age. This incomplete outgassing is unlikely to be the cause of the older carbonate ages.

This analysis indicates that 14C dating of carbonate associated with lime burials and habitational fires, yields reliable ages. Unfortunately, the range of 13C values from -12 to -21 o/oo is largely the result of kinetic effects. Thus, anomalous ages cannot be detected always using 13C ratios. Inhumation lime dating, on average, yields much better results than mortar dating, but a criterion that always distinguishes normal material from anomalous is not yet available.

The accurate 14C dating of mortar depends on the extent of contamination by ancient carbon resulting from the use of calcareous sands and/or incompletely kiln-fired limestone (**6-8**). As these factors depend on geographic locations, a few Mallorcan samples were measured. Stucco from Pollentia Roman Republican building at Alcudia, dating from about 30 B.C. has a very anomalous age of 8,000 B.P. (**Y-2668**). On the other hand, stucco from Santa Catalina de Sena excavation, dated through its association with Moorish pottery at about 1000 to 1200 A.D. has an age of 14C, (**950 b.p.**) ± 100 years (**Y-2671**). Evidently, the lime making techniques of these two cultures were quite different.

The 14C age of lime carbonate at the Talayotic Son Oms site was also investigated. It was formed after the local limestone bedrock was decomposed by overlaying fires of Talayotic origin. There a completely anomalous age of 20,750 years was obtained (**QL-19**) but in that instance the 13C ratio of -5 o/oo clearly showed that incomplete degassing of the original bedrock limestone caused the discrepancy.

Discussion

The duration of the Post Talayotic Period becomes clear in the light of this new information supplied by absolute dating methods when applied to the lime inhumation stratigraphies' of the Matge and Muertos Gallard cemeteries. At the present time, several other sites are undergoing study while further information is being sought in both the former stations. There is no reason to believe that new datings will not fall within the current framework; certainly filling in some of the gaps and adding to our overall knowledge of the custom and the period during which it was practiced.

The most interesting feature of the framework of conventional14C dates and those of 14C carbonate dating is the evidence it presents for the sub-division of the 700 to 800 year Post Talayotic Iron Age Period, the period into Initial, Early, Middle and Late phases. It also offers vital information as to the transitional period of the Late Talayotic Bronze Age, establishing a tentative date of 800 to 850 b.c. for this stage.

It should be pointed out that the collection of test samples was designed to give maximum information about critical sectors and about the conditions present in the stratigraphies. Not every area or level could be tested for practical purposes, although numerous samples have been collected for future use. These will be analysed as the need arises.

Recently an extensive series of these samples collected for this purpose have been sent to Belgium for analysis by M. Strydonck as part of a research programme designed for the dating of ancient motar and lime. Unfortunately, none of this series of results is available at the current writing, but will be available in a future edition of this Balearic dating inventory.

Meanwhile, sample collection continues to be systematically carried out, and as often as possible done so in levels and sections where artefacts of recognisable types have been well preserved. In this way, the project data serves a twofold purpose; that of dating both the critical strata and the artefacts they contain.

Of the best preserved artefacts from the quicklime contexts, the most useful have been those made of iron. Where pottery has survived in any quantity at all, it has done so in the peripheral areas of the cemeteries, though the occasional pieces are present in the lime beds themselves. It appears, however, that this custom of burying the pottery apart from the bodies was frequently the case during most of the Post Talayotic Iron age. It is either this or the pottery that normally would have been present in the quicklime has been destroyed by the slaking action of the lime, and as a result those pieces that are present on the perimeters of the beds escaped destruction only by their location of the cemetery perimeter.

Apart from the artefacts, most of the information comes from the stratigraphy of the deposits themselves, and the interested reader is referred to the detailed plans and sections of the author's doctoral dissertation for a full documentation of this; although a stratigraphic section and some discussion of these pertinent levels are given the reader in appropriate sections (**Parts III and IV**) of the current publication.

The information as a whole has given us an excellent insight into the Post Talayotic Period Iron Age , as well as help to establish parallels and make comparisons with Continental Iron Age cultures. Although this new gained insight for the moment is still not as comprehensive as would be desired in its interpretation of the period. But then again such is never quite the case. In this sense, the results as offered are admittedly fragmentary and incomplete. The objective here is <u>simply</u> to offer an up-to-date account which supports the proposed sub-division of the Post Talayotic Iron Age in more absolute terms than has been so far possible. At the same time, they illustrate and emphasise the complexity of the local problems regarding the Late Talayotic and Post Talayotic Periods, as well as some of those pertaining to insular studies in general.

In the preceding sections of this particular paper, we have examined a few of the problems presented by the change-over from the **parent** Talayotic Bronze Age culture to the Post Talayotic Iron Age, chiefly through the study of changing burial customs and technology. We have seen some of the implications brought about by those changes, and a little of the range of research and study carried out during the last decade towards their solution. However, there are still other possibilities offered us in the stratigraphies themselves which can further enlarge our understanding of the period as well as that of other periods.

II

PRIMARY, SECONDARY and TERTIARY RESEARCH SITES

The research stations examined in this publication can be divided into three main categories. Only the first and second categories (**primary and secondary sites**) are considered in detail by order of rank in terms of their duration of occupation, which in the cases of the **primary** sites span several periods. The third category (**tertiary sites**) are dealt with in less detail or in reference only; despite the fact that at least two of these tertiary sites have important radiocarbon series. This really should qualify them as primary or secondary sites on this basis as they are of considerable importance and interest to the overall interpretation of Balearic Talayotic open-air settlement patterns. However, the down grading of these sites is done on the qualification of their occupational duration and for no other reason.

In the first of the two cases, the Son Fornes Talayotic Settlement (**Gasull, Lull and Sanahuja 1984**), the site's full duration of occupation is not known, appearing at the present stage of excavation to extend over only one prehistoric period. In the second case, while clearly like Son Fornes in having a single period duration of occupation, the Settlement Complex of Torralba den Salort (**Fernandez-Miranda and Valdren in press**) also appears to have had an even earlier one; although this earlier level of occupation is not as yet satisfactorily defined. In any case, these two **tertiary** sites are included and briefly described along with others in due course, and their interaction with the **primary** and **secondary** sites examined, if only in a rudimentary way.

The actual types of sites studied in this paper, regardless of their rank as to duration of occupation, are indeed varied and differentiated by the many characteristics and conditions found in them. These differences exist in all three categories of sites, but are especially evident regarding the **primary** research sites examined. For instance, the **primary** sites differ from one another in physical characteristics (**geology**), as well as situation (**geography**) and the kinds of stratigraphy (**deposition**) present in them; as well as the class and range of materials (**artefacts**) they contain.

However, the common characteristics which they share and the qualifications which designate them as **primary** sites are their known and long duration of occupation as well as a consideration of the particular or special stratigraphical conditions and material contents found in them; characteristics and conditions which as another requisite have been exceptionally well delineated and documented by radiocarbon dating.

The **secondary** sites are those in which characteristics, conditions and materials, whether similar of not, are lesser in degree; although all of these sites too have some radiocarbon documentation as well as correlative evidence that is pertinent within the framework of the Balearic chronological /chronometric sequence; or are otherwise notable in some way. For example Muertos Gallard and Marroig as secondary sites are not only geographically adjacent to one another, but geographically form part of the chain of research sites, including the primary ones, along Mallorca's North Coast (**Figure**).

In summary, it is the special nature of all three categories of sites that is accountable for their having been the subjects of the current research in the first place. It is also one of the qualifications that makes them of value as sources for the interpretation of local prehistoric events, as well as being a general contribution to European Prehistory. At the same time, as they stand at present, they are representative of the new generation of Balearic investigation.

In fact, the Balearic chronolological/chronometric sequence, on the whole, is a representative prehistorical overview of a relatively small and selected geographic area, compared to the wider geographic contexts of mainland continental regions. However, these limitations and the other insularity factors they naturally impose become less important when we closely examine the wide variety of characteristics and conditions, as well as the range and grade of evidence both physical and otherwise, which are offered us by intensified, insular research. And simultaneously, pointing out to us the advantages to be gained and expected from geographically limited areas, which have undergone long-term, interdisciplinary investigations.

The **primary** sites, which we are about to examine are three in all: (1) the **Cave of Son Muleta (SM)**, (2) the **Rock Shelter of Son Matge (ABSM)** and (3) the open-air settlement complex of **Ferrandell-Oleza (SFO-OS and YS)**. These three primary sites along with two of the secondary ones discussed subsequently form a chain of archaeological stations along a twenty kilometer sector of the north coastal mountain range of the island (**Figure 11**), and form a fairly representational as well as varied geographical cross-section of this section of the coast.

THE PRIMARY RESEARCH SITES

«««««««««««««««««««»»»»»»»»»»»»»»»»»»»»»»»

The Cave of Son Muleta (1. A. SM)(6°,20'E; 39°,49'N)

Introduction

‹‹‹‹‹‹‹‹‹››››››››

The Muleta cave deposit consists of an unbroken stratigraphical prehistoric sequence, dating for well over 100,000 years. At the same time, '**deep**' physical stratigraphies like Muleta's are just not frequently found in the Balearics, being rare by standards anywhere. On this basis- as pointed out earlier in the introduction- when linked up with the '**deep**' stratigraphy of the Matge deposit, it forms an integral part of a veritable '**yardstick**' of prehistoric events both of **ecological** as well as **cultural** importance. Therefore, they should not be broken down and categorised solely on the basis of the scientific disciplines involved or the various kinds of data and information such complicated stratigraphical sequences offer, but only in order to describe the succession of events and materials found in such deep stratigraphies.

A division of the primary research sites into strictly palaeontological and archaeological contexts is still less reconcilable when one further considers the fact that this **eco-palaeontological** nature of the Muleta stratigraphy, along with its own archaeological contexts, also links itself with the equally important stratigraphic cultural sequence at the Matge rock shelter and Ferrandell-Oleza Settlement Complex. In so doing, the three sites together form an overlapping entity, creating a still more important and detailed 'yardstick' for Balearic prehistory. This overlap between sites is demonstratable in Figure 1.

Besides, the Muleta **Presettlement Period (PS)** horizons with their ***Myotragus balearicus*** materials contain a wealth of palaeontological data and information as well as materials which become particularly relevant when we come to study the **Early Settlement Period (ESP)(Part III)** and the activities of the islands' early settlers. As these ESP context do not occur before we reach the top-most layers of the PSP stratigraphy in the Muleta deposit, we must first consider the site's combined physical stratigraphy in terms of chronological duration (**chronometric and relative**), and secondly the physical evidence, its description and what we can suggest from it concerning nature, evolution and the general ecological conditions present. This is done elsewhere through the medium of such considerations as sedimentological and other studies, for example the osteological characteristics of the ***Myotragus balearicus*** as to its morphological peculiarities and function along with other aspects such as sexual dimorphism (**e.g. Appendices 1D & 1E Waldren 1982**). Most of these topics are examined briefly in the appropriate sections of this present publication as they seem applicable and particularly relevant. Equally, as a case of continuity and immediacy, a physical description of the Muleta cave is given below. This is included with the cave's stratigraphical discussion, along with an illustrative example of one especially pertinent stratigraphy, that of 'X' Sector of the cave (**Figure 3**). There are other cave sectors which are equally important, but 'X' Sector best exemplifies the origin and age of the cave's deposit.

Therefore before we study the Muleta stratigraphies, their absolute chronology and the materials representing the Presettlement Period, we shall first examine the geographic situation as well as the geological structure of the cave. For it has been these physical characteristics above all that have been responsible for the preservation of the deposit's materials and also for the success of the dating methods used on them. While the location of the site is found in an area which has been well protected over the many millennia before its discovery, it has been the physical structure of the cave itself which has been the contributing factor for that preservation.

Anyone familiar with caves comes to know that their physical structure varies a great deal and that each cave is different and controlled by an individual set of internal conditions in regard to their atmosphere and thermal ambience. However, it is generally accepted and expected, at the same time, that materials will be well preserved in cave situations. In this respect, the Muleta cave is no exception. Nevertheless, the Muleta cave is highly individual in that its structural characteristics were such that the ambient conditions in the cave have resulted in what is best defined as maximum preservation of its materials.

Of equal importance are the cave characteristics which led to the deposit of its stratigraphy in the first place. In the case of Muleta, its function as a natural animal trap lies in its unusual formation and configuration. In short, we have in the Muleta cave a special combination of physical (**geological**) characteristics (**constructional and depositorial**) which having been brought together and into play made it into an excellent and reliable reservoir for the application of a number of different types of studies; e.g. the present chronometric survey and the methods described in it.

The Geographic Location and Geological Structure of the Muleta Cave

«««»»

The Muleta cave is located 5 kilometers from the small mountain village of Deya, population 600, and 5 kilometers from the market town of Soller, population 10,000. The north coast road, Palma-Valldemossa-Deya-Soller, is one of the two access roads to this coast . The other is the longer route to Soller, Palma-Soller, by way of the Coll de Soller with its present-day 88 hairpin turns. The north coast road was certainly the easier of the two access roads even in ancient times, though other traffic must have passed via the Coll de Soller route. Besides being the easiest route into the mountains, the north coast road would have given complete access to the coast between Valldemossa and Soller, the route along which the author has discovered all of the primary and secondary prehistoric sites, including the rock shelter of **Muertos Gallard** and the cave of **Son Marroig (inventoried AMG and SMRG, respectively)** each of which are examined here. These latter two are the secondary research sites situated between Ferrandell-Oleza on the south and Matge and Muleta to the north, and are essential links, as pointed out earlier, in the chain of archaeological stations used in this research (**Figure 10**).

The outcrop of Jurassic limestone containing the Muleta cave system lies below and on the seaward side of the 5 km road marker on the coast road between Deya and Soller. There is a small side road at this point which descends to the urbanised part of the coast of Deya and Muleta Vignon. The area is a promontory of land known as Muleta Gran. The outcrop is situated at an elevation of 180 meters, directly overlooking the sea, rising steeply from the flat farmland which forms part of the promontory of this section. The flat lands form the only level region along this particular stretch of coast until the valley of Soller. From this point on, the coastline east of Soller is made up of some of Mallorca's most precipitous cliff sections. The flat area of Muleta would have been ideally suited for both man and animal in ancient times. There is evidence of an old torrent bed immediately in front of the outcrop which is now filled in and terraced to make available more land to agriculture. These terraces end abruptly to the north, exposing the original torrent's steep descent to the sea directly below.

As a solution cave in Miocene limestone, Muleta is not what one would expect in regard to its sedimentary or fossil accumulation. Made up as it is of two horizontal, right angled and two levelled systems that are connected by a single narrow vertical chimney, the system functioned as a natural animal trap whose internal environment has been exceptionally constant during most of the Holocene. This is reflected in the excellent preservation of materials and can also be demonstrated from results obtained in the various analyses where constant temperatures are either requisites or advantageous. Records kept by the author of the temperatures in the cave for all seasons and weather conditions over the period of 5 years show a yearly mean temperature of 19.5 degrees centigrade, with summer temperatures not exceeding 20 degrees centigrade and winter ones not falling below 19 degrees centigrade.

There are two openings to the cave system from outside, which are (**a**) the main entrance in the form of a key-hole shaped mouth (48 cms x 125 cms) and (**b**) the inner mouth of a vertical chimney to the lower cave system (48 cms in diameter) (**Plates 1:1, 2:1**) (**Plan 1A and 1B; Waldren 1982**). It is their small size which has been responsible for the good state of preservation of the cave's contents and the lack of variation in its ambient temperatures. Their size and nature also explain the success of the cave as an animal trap.

The two chambers composing the cave system are an upper one aligned roughly north-south and a lower one, east-west. The upper chamber, reached from the outside by way of the key-hole entrance, is 7m long x approximately 2m wide with varying heights of 1m to 3m (**see Plan 1 and 1B; Waldren 1982**). This lower cave lies at the base of the narrow chimney at the rear of the upper chamber and slopes at a 45 degree angle (**east-west**). This lower cave is about 6m long x 1.5m wide x 5.75m high as it is today. These caves were formed in a '**karst**' fractured limestone, the original fracture zone being enlarged by solution action prior to accumulation of the sediments.

Stratigraphical Contexts and Dating of the Muleta Deposit: Summary

Over many millennia, the Muleta cave became filled by a gradual build up of faunal remains and cave sediments of different kinds and origins and continuing until the present. The amount of bone material present is remarkable; and the accumulation of the cave of approximately 2000 examples of ***Myotragus balearicus***, innumerable microfauna of the same period, skeletal remains of early human inhabitants and finally modern fauna is unprecedented in the Balearics. One stratigraphical profile in 'X' Sector of the cave (**Figure 3**) used as a stratigraphical control area

Table. 10

	Human Occupation			
	1.	2.	3.	4.
Gramineae	18	51	54	
Cyperaceae		4	13	
Triticum type		2	1	
Typha		1		
Lillinceae		2		
Asphodelus		2		
Taraxacum	1	3	4	
Ligalifloreae	1		3	
High spine compos.	1	7		
Low spine compos.	58		1	
Artemisia		1	2	
Plantago		5	5	
Caryophytacese				
Leguminoseae		1	1	
Chenopodiaceae	1		1	
cf. Plumbaganaceae		1		
cf. Salicighia	2	1	2	
Umbelliferae		1	1	
cf. Cruciferae		2	1	
cf. Lepidium		3		
Jasione				
Pinus	5	1	1	
Cupressaceae	1			
Fraxinus	4	2		
cf. Tamarix	3	2		
Quercux		1	2	
cf. Juglans		2	1	
Corylus			1	
Betulaceae			1	
Alnus			1	
Acer			1	
EAP	13	8	7	
Filcoles				
monolete type A	1	1	1	
monolete type B			1	
polypodium		1		
trilete type A		1		
trilete type B				
tetrad type A		1	1	
Unknown	4	1	1	
Microns	100	100	100	

Cave of Son Muleta
Soller, Mallorca

No. 1 - Sector Z, 300cms
No. 2 - Sector F, 150cms
No. 3 - Sector X, 50cms
No. 4 - Sector Z, 250cms

Analyses: Geochronological Laboratory, University of Arizona, Tuscon, Arizona

TABLE

10

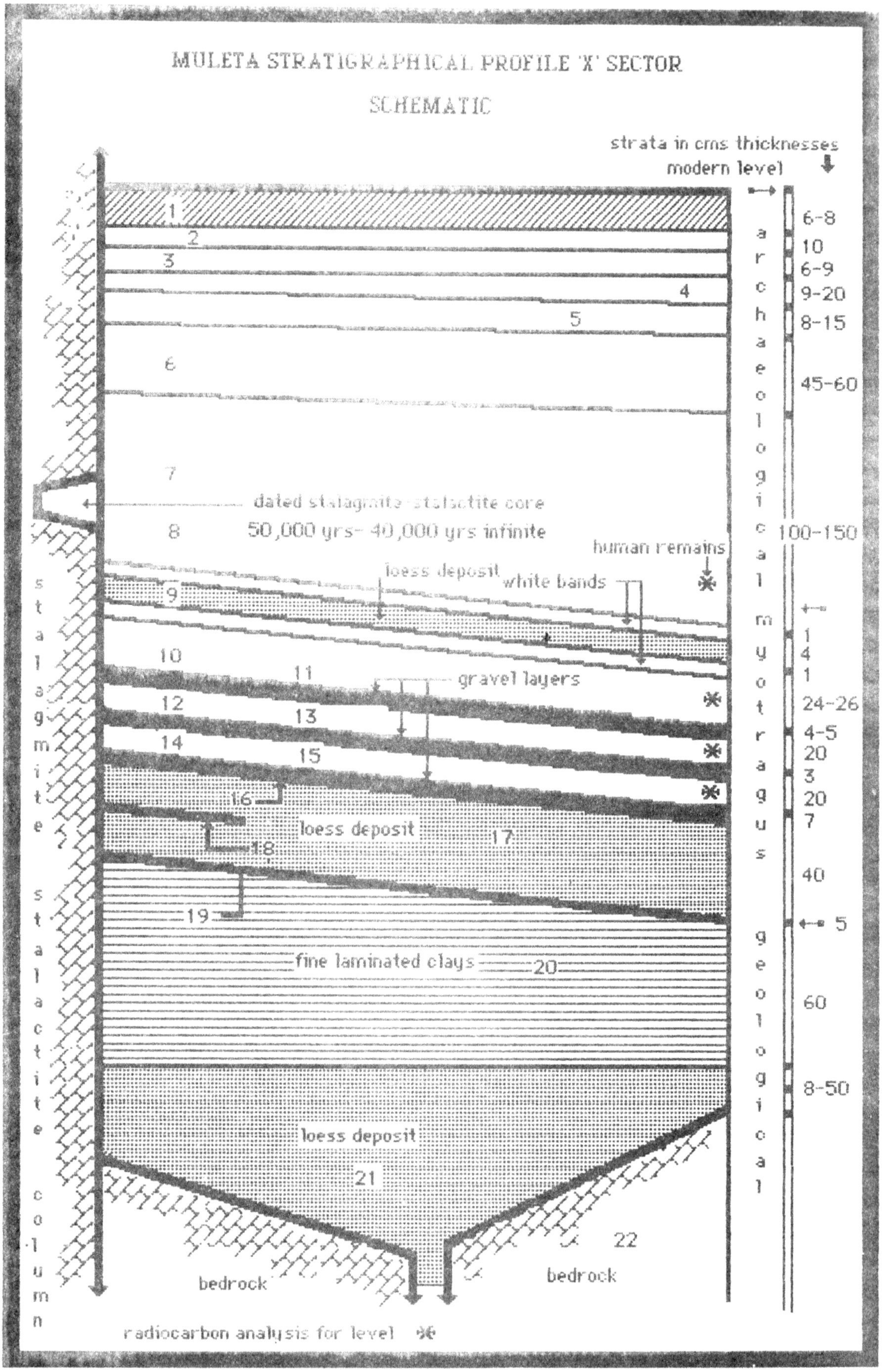
MULETA STRATIGRAPHICAL PROFILE 'X' SECTOR
SCHEMATIC
strata in cms thicknesses
modern level
1
2
3
4
5
6
7
dated stalagmite-stalactite core
8
50,000 yrs- 40,000 yrs infinite
human remains
loess deposit
white bands
9
10
11
gravel layers
12
13
14
15
16
loess deposit
17
18
19
fine laminated clays
20
loess deposit
21
22
bedrock
bedrock
stalagmite
stalactite
column
archaeological
myotragus
geological
6-8
10
6-9
9-20
8-15
45-60
100-150
1
4
1
24-26
4-5
20
3
20
7
40
5
60
8-50
radiocarbon analysis for level ✻

FIGURE

3

produced 22 levels of accumulation; representing approximately 200,000 years beginning with what is believed to be wind blown '**loess**' (**Munsell 7.5YR, 4/8**) of the Riss Glaciation (**see section below**).

The cave, when discovered in 1962, was blocked with brush, which had been placed in the cave's entrance to prevent present-day grazing sheep from wandering into the cave in search of shelter. During the period of work at the site the local shepherd has always been most concerned with re-sealing the cave whenever operations are not in progress, against accident to his flock. This entrance to the Muleta cave from the exterior on first impression gives very little idea of what it contains, as it is only just large enough for a human to crawl through head first. An animal however would have no difficulty, especially one like ***Myotragus balearicus*** which stood only as high as a medium sized dog and weight about 13 to 14 kilos. Once in the small entrance of Muleta, human or animal visitors would have been in jeopardy as soon as they passed beyond the periphery of light, as the chamber ends in the vertical chimney which gives access to the lower chamber of the system (**Plan 1A, 1B and 1C and Plate 1:1; Waldren 1982**).

The cave filling was accordingly formed in the above manner, mainly in the lower part of the system, slowly building over the millennia. The deposit of bones represents animals, and in due course humans also, who evidently, sought shelter and accidentally became trapped. Indeed, the deposit must be classed as a natural animal trap rather than an occupational site. The rate of its accumulation of fossil remains as well as sediments are studied in a appropriate section of the author's thesis (**Appendix 1B; Waldren 1982**).

The Presettlement Period (Strata Inferior to Stratum 7)

In totality, the absolute dating of the Muleta deposits that represent the Presettlement Period in the various stratigraphies (O, EF, CD, AB, X and Z sectors, each studied below) combined cover a time from about 80,000 b.p., according to palaeomagnetic dating (**SM 7**) to 7000 b.c. by radiocarbon and racemisation of aspartic acids analysis (**Appendix 1A**). The results of the 14C and RAA assays on ***Myotragus*** bones from the presettlement levels have provided a chain of 16 dates from about 32,000 b.p. to about 7000 b.p., an average of 1 dating per 1550 years of accumulation of the cave's filling. Agreement between the various sectors is generally very good. In most sector stratigraphies, it has been possible to assign estimated dates to certain strata which are 'sandwiched' between two levels for which actual chronometric dates exist, e.g. as has been done in the study of 'Z' sector cave stratigraphy (**section 4.3, Vol. I, Valdren 1982**). It has also been possible to demonstrate the order in which accumulation of the deposit in the various sectors took place, concurrently in some cases and consecutively in others, thus giving us another type of chronological interpretation. The geochemical interpretation of the sediments suggests that they began to be deposited as early as the Riss Glaciation, perhaps since circa 230,000 b.p., while the main bone accumulations began as early as 32,000 b.p. in a middle phase of the Wurmian Glaciation, continuing into what is dated in northern Europe as the Atlantic Phase of the Post Glacial Period or Holocene.

The Early Settlement Period (Strata 7 to 6)

The evidence directly establishing the existence of an Early Settlement Period at Muleta consists of **(a)** human skeletal remains that represent several individuals, in spite of the absence of some of the skeletal components, and **(b)** a few artefacts (flint flakes and a number of bone needles). While the level from which these originate (**Stratum 2**) (**150cm to 175cms**) is well documented by 14C, the evidence has only recently become sufficient for us to do more than note the possibility of such a period in the prehistoric sequence in the Balearics. As we will see, it has taken the correlating evidence of the Matge rock shelter and the consequently early radiocarbon dating to place the existence of the Early Settlement Period beyond doubt.

Chronometric dating of the Early Settlement Period comes only from Stratum 7 in the Muleta deposit, which contained the anthropological and palaeontological samples used in the age determinations; however, the dates are remarkably in accord with those more recently established at Matge. The materials used in the first radiocarbon analysis of human bone were chosen from a very small amount of existent materials and hence only small test samples were used in the initial analysis, but as we will see the results were not at all satisfactory. The results of these,: 8448 b.p. (**6498 b.c.**) ± 1180 years (**SM 5**) and: 10,685 b.p. (**8735 b.c.**) ± 3517 years (**SM 59**) were obtained from 86 grms of ***Myotragus*** bone and 39.5 grams of human bone respectively. Naturally, dates with such large statistical errors were not much use, except in so far as they gave some rough idea of the specimen's likely age. It was recommended by the laboratory concerned that larger test materials should be submitted for analysis; bone specimens, both human and animal, weighing 500 grms each were at once submitted. The results are those presently used for the Early Settlement Period: 5935 b.p. (**3985 b.c.**)± 109 years (**SM 54**) for

Table Inventory of Human Bones: Various Sectors 'O", 'EF', 'CD', 'AB' and 'X', 150-175cms.

No.	Bone	No.	Bone
001.	Patella (R)	049.	Phalange, Manus, II (R)
002.	Patella (R)	050.	Phalange, Manus, II (R)
003.	Metacarpal III (L)	051.	Phalange, Manus, II (R)
004.	Metacarpal III (L)	052.	Phalange, Manus, II (R)
005.	Metacarpal III (L)	053.	Phalange, Manus, II (?)
006.	Metacarpal V (R)	054.	Phalange
007.	Metacarpal IV (R)	055.	Phalange, Manus, I
008.	Metacarpal II (R)	056.	Phalange, Manus, I
009.	Metacarpal I (R)	057.	Phalange, Manus, I
010.	Metacarpal IV (L)	058.	Phalange, Manus, I
011.	Metacarpal I (L?)	059.	Phalange, Manus, I
012.	Phalange, Manus, Y, I (L)	060.	Phalange, Pes, I
013.	Phalange, Manus, Y, I (R)	061.	Phalange, Pes, I
014.	Capitate (R)	062.	Phalange, Pes, I
015.	Capitate (L)	063.	Phalange, Pes, I
016.	Scaphoid (R)	064.	Phalange, Pes, I
017.	Lunate (L)	065.	Phalange, Pes, I
018.	Trapezoid (L)	066.	Phalange, Pes, I
019.	Metatarsal V (R)	067.	Phalange, Pes, Y, I
020.	Tibia (L)	068.	Phalange, Pes, II
021.	Talus (R)	069.	Phalange, Pes, II
022.	Cuneiforme II (L)	070.	Phalange, Pes, III
023.	Tibia (L)	071.	Phalange, Manus, Distal
024.	Tibia (R)	072.	Phalange, Manus, Distal
025.	Radius (R)	073.	Phalange, Manus, Distal
026.	Metatarsal I (L)	074.	Phalange, Manus, Distal
027.	Trapezoid (L)	075.	Phalange, Manus, Distal
028.	Tibia (L)	076.	Phalange, Manus, Distal
029.	Calcaneum (R)	077.	Phalange, Manus, Distal
030.	Calcaneum (L)	078.	Metatarsal, I (R)
031.	Vertebrae, Thoracic	079.	Metatarsal, II (R)
032.	Vertebrae, Atlas	080.	Phalange, Manus, I, Distal
033.	Vertebrae, Cervical	081.	Phalange, Manus, II (L)
034.	Phalange, Pes, Y, I	082.	Phalange, Manus, II (L)
035.	Phalange, Manus, II (R)	083.	Phalange, Manus, II (L)
036.	Phalange, Pes, I	084.	Phalange, Manus, II (L)
037.	Phalange, Manus, II (R)	085.	Phalange, Manus, II (L)
038.	Metatarsal II (R)	086.	Vertebrae, Lumbar
039.	Metacarpal, IV (R)	087.	Molar
040.	Metacarpal, IV, (L)	088.	Molar
041.	Metatarsal, IV (L)	089.	Molar
042.	Metatarsal, IV (R)	090.	Molar
043.	Metacarpal, I (L)	091.	Molar
044.	Metacarpal, I (L)	092.	Molar
045.	Hamate (R)	093.	Molar
046.	Vertebrae, Cervical	094.	Molar
047.	Metatarsal, I, Y, (R)	095.	Premolar
048.	Phalange, Manus, II (L)	096.	Premolar

TABLE

11

Table
Inventory of Myotragus Bones: the Rock Shelter of Son Matge

No.	Bone	Sector/Level	Description
001.	Horn Core, L or R	44-42, 475cms	Fragment, Tip
002.	Horn Core, R	44-42, 475cms	Fragment, Base
003.	Vertebrae, Lumbar	44-42, 475cms	Damaged
004.	Vertebrae, Lumbar	44-42, 475cms	Damaged
005.	Femur, R	44-42, 475cms	Damaged
006.	Humerus, R, Proximal	44-42, 475cms	Damaged
007.	Mandible, R	44-42, 375cms	Burned
008.	Femur, R	44-42, 475cms	Fragment, Scarred
009.	Rib, Proximal End	44-42, 475cms	Butcher Scarred
010.	Ulna, R	44-42, 475cms	Butcher Scarred
011.	Humerus, L	42, 350cms	Complete
012.	Scapula, R, Proximal	42, 350cms	Butcher Scarred
013.	Femur, L	42, 350cm	Burned
014.	Ulna/Radius, R	42, 350cms	Burned
015.	Metacarpal, L	42, 350cms	Burned
016.	Vertebrae, Thoracic	44-43, 475cms	Damaged
017.	Molar, Maxilla	44-43, 475cms	Complete
018.	Femur, R, Distal	44-43, 475cms	Butcher Scarred
019.	Scapula, L	44-43, 475cms	Butcher Scarred
020.	Scapula, L	44-43, 475cms	Butcher Scarred
021.	Pelvis, R	44-42, 475cms	Butcher Scarred
022.	Pelvis, R	44-43, 475cms	Butcher Scarred
023.	Pelvis, R	44-42, 475cms	Butcher Scarred
024.	Pelvis, L	44-42, 475cms	Butcher Scarred
025.	Pelvis, L	44-42, 475cms	Butcher Scarred
026.	Pelvis, L	44-42, 475cms	Butcher Scarred
027.	Humerus, L	30-26, 160cms	Damaged
028.	Metacárpal, L	30-26, 160cms	Complete
029	Maxilla, P	30-26, 160cms	Fragment
030.	Maxilla, R	30-26, 160cms	Broken
031.	Mandible, L	30-26, 160cms	Broken
032.	Mandible, R	30-26, 160cms	Broken
033.	Radius/Ulna, R	42, 350cms	Complete
034.	Radius/Ulna, L	42, 350cms	Complete
035.	Radius/Ulna, R	42, 475cms	Butcher Scarred
036.	Rib, L, Distal	42, 350cms	Damaged
037.	Rib, L, Distal	42, 350cms	Damaged
038.	Rib, R, Distal	42, 350cms	Butcher Scarred
039.	Rib, L	42, 350cms	Demaged
040.	Rib, R, Distal	42, 350cms	Damaged
041.	Rib, R, Distal	42, 350cms	Butcher Scarred
042.	Rib, L, Dista	42, 350cms	Burnt
043.	Tibia, L, Distal	42, 350cms	Pathological
044.	Cranium	42, 190cms	Broken Right Horn Core
045.	Cranium	42, 350cms	Burnt, V Trimmed Horn Cores
046	Cranium	42, 350cms	Broken for Brain Extraction
047.	Cranium	42, 350cms	Both V Trimmed Horn Cores
048.	Cranium	42, 350cms	Both V Trimmed Horn Cores
049.	Horn Core	44-42, 350 cms	V Trimmed
050.	Horn Core	44-42, 475cms	V Trimmed

TABLE

12

the human bones themselves and 7135 b.p. (**5185 b.c.**) ± 80 years (**SM 56**) and 8570 b.p. (**6620 b.c.**) ± 350 years (**SM 57**) for the ***Myotragus*** remains associated with the evidence of man's presence.

Depositional details and a discussion of these materials are found in the sections concerning the pertinent level (**Stratum 7**) of the various stratigraphies (**0, EF, CD, etc.**) and are also found in the inventory of the human remains (**Table 11**). However individual descriptions of all the cave deposit's stratigraphies are not possible in the present publication but are covered in great detail elsewhere (**Valdren 1982, Text and Appendix 1A**).

The Pretalayotic Period (Strata 6 to 4)

The evidence for this period and the subsequent Talayotic Period is based on a small series of habitational levels (two for the Pretalayotic Period and one for the Talayotic Period), representing short spells of occupation of the cave, probably as a transient camp for shepherds or a small family unit. Pottery evidence is also present for both the Pretalayotic and Talayotic Periods, dating from circa 3910 b.p. (**1960 b.c.**)± 120 years (**SM.47**) and 2765 b.p. (**815 b.c**).± 120 years (**SM.27**) respectively.

Details discussing these levels and materials are found in the text and in the sections dealing with the different stratigraphies below.

The Talayotic and Post Talayotic Periods (Strata 4 to 0)

The Talayotic and Post Talayotic levels of the Muleta cave deposit like those of the Pretalayotic Period, are not very extensive. Like those of the Pretalayotic Period, they represent short occupational phases by a small number of people, probably transient shepherds; as we have seen the small size of the Muleta cave could not have possibly sheltered more than a very small number of individuals, and probably only a single shepherd using the site in bad weather. The preceding level in the Muleta sequence (**Stratum 5**) we have seen to have been in process of formation circa 1960 b.c. We will note that if the radiocarbon analyses are to be believed, the cave underwent a very long period of abandonment of over 1000 years, during which time it may actually have been sealed off. At any rate, there is no sign either of animals trapped in the lower system nor human occupation in the upper area until circa 815 b.c., when the cave was again used as a transient camp.

The Post Talayotic Period in Muleta (**Strata Z to O**) is not a human occupational level, but the period is represented by a palaeontological level containing goat remains of animals which had fallen into the lower chamber as ***Myotragus*** had done. Radiocarbon analysis reading on bones of these specimens were circa 2180 b.p. (**230 b.c.**)± 115 years (**SM.4**).

With the completion of this summary of the general stratigraphy and chronology, we can now begin the detailed description of the individual levels of the different stratigraphies at Muleta (**e.g. 'X', 'O', 'Z', 'EF' etc**). While there is a certain homogeneous nature to the Muleta deposit that is shared by the various stratigraphies, there are distinctly different contextual and stratigraphical situations as well as deposit characteristics which must be individually described. This is the author's reason for presenting the different sector's stratigraphic profiles and a more schematic description and presentation. The accumulation and other depositorial factors in these different sectors are very determining ones in the final interpretation of the deposit, especially as regards to the chronology of the deepest areas of these sectors, where dating is either out of range or in the far range of radiocarbon method.

A Summary of the Methodology Used in the Excavation and Research

The Excavation and Earth Removal

While the elements of the cave's architecture- small chambers and entrances- were the controlling factors in the accumulation of the deposit, they also are responsible for the methods used in excavation. The confined working conditions throughout the cave system were extremely cramped. In order to facilitate the removal of the excavated earth and specimens, in particular, their removal up the vertical chimney, through its mouth and subsequently outside the cave via the key-hole entrance, a system of pulleys and transport cables had to be installed (**Waldren 1982, Appendix 1B**) to cope with these otherwise difficult bottle necks. Light weight plastic buckets were used for the transport of earth and specimens on the pulleys.

The specimens, once excavated, were packaged in plastic specimen bags in the cave with identification cards following the classification system outlined in section 2.1 (**Waldren 1982**), and were then transported up the chimney and through the chimney mouth. Here, the buckets were handed through the upper chamber and attached to a steel cable and runner-pulley system leading out of the key-hole entrance, down to the work areas. Here, the buckets were emptied onto clean plastic sheets assigned to the sector and level being excavated, to await further processing.

The Hydraulic Washing System

An area immediately in front of the cave entrance was equipped with a specially devised soil washing system consisting of a washstand, racks and hoses for processing the cave earth by washing it through 7 different screens which trapped the most miniscule bone components and at the same time graded their containing matrix into stone, gravel, sands and clays. This hydraulic separation system for the cave earths is the subject of a more detailed description in the appendices, and the drawings are given for the field equipment which was designed by the author (**Appendix 1B**).

Earth Storage and Weighing

The excavated earths were first weighed before washing, each of the individual buckets of earth's weight being recorded. The various residues that were graded in separation were weighed after washing and sun drying, and the differences in weights of the unwashed and washed soil was recorded as washed out clays. Once the separated graded residues had been sun dried, they were packaged in double, stout plastic bags and labelled as to origin, in preparation for bone specimen extractions. In this way, well over 90% of the entire deposit was excavated, washed, statistically recorded and had all remains removed, even the smallest fragments (**Plates 7:1 and 7:2, Waldren 1982**).

The Rock Shelter of Son Matge (1, B, ABSM)(6°, 25'E; 39°,35'N)

Introduction

Like the Muleta cave, the Matge rock shelter contains an unprecedented and unbroken stratigraphical sequence. The main difference between the two sites - apart from the physical ones - are the chronological durations of the two deposits; while the Muleta deposit consists of a palaeontological sequence of levels and materials of some [illegible]00,000 years duration, the Matge stratigraphy is made up of some 5,000 years of cultural accumulation. When we link the two stratigraphical sequences together, the Matge sequence not only forms an extension to the already long Muleta one, but together forms a still more unusual chronometric **'yardstick'** especially since Matge also has had very extensive radiocarbon documentation of its many different cultural horizons.

Equally important is the fact that the Matge stratigraphy supplies us with supportive evidence related to the association of man and ***Myotragus*** first found in the Muleta deposit (**Waldren 1968; Rosselló Bordoy and Waldren 1973; Fernandez-Miranda and Waldren 1976**); supportive evidence which greatly enlarges various aspects of that relationship, while at the same time permitting us to distinguish an Early Settlement Period, circa 5000-3000 b.c. Furthermore, whereas the Muleta cave produced evidence of man in the Fourth Millennium b.c., the Matge site extends that early presence of man in the islands to a period still earlier in the Fifth Millennium, b.c. along with giving us many more details of man's activities at that time. As we will see below, these activities can be clearly described from the stratigraphical remains and information for a continuum of about 5000 years, to embrace all the Balearic cultural prehistoric sequence. Hence Matge, its situation, stratigraphy, materials and function will occupy and concern the subsequentsections of this text.

The physical differences between those two highly important sites are needless to say considerable; Matge being huge in comparison to Muleta and in turn having been used for multiple purposes throughout time. While the Muleta stratigraphy was complicated in its sequence of deposition as a natural deposit, the Matge accumulation is perhaps still more complicated, apart from its size, because of the natural erosion of the face of the rock shelter and the accumulation of man and the uses to which he put the area. Therefore an account of the various uses and conditions found in the shelter's various areas must be considered, if the reader is to understand the sequence of events as they occurred during the various prehistoric periods present in the Matge deposit.

The Geographic Location and Geological Structure of the Matge Shelter

Matge (**see map, Figure 19**) is located 13.5 kilometers from the Balearic capital of Palma (**pop. 300,000**) and 4 kilometers from the mountain village of Valldemosa (**pop. 850**) in the narrowest section of the north coast road known as S'Estret or the narrow place where the road begins its steep ascent into the mountains. Here the north coast road starts to wind itself through the cleft in the mountains, forming a sharp 'V' which narrows to about 20 meters. At this point in the pass there is space only for the two lane highway and a steep sided stream which makes its way down into the plains. This stream would have been an ancient water source, even as it is a modern one. In prehistoric times anyone controlling this narrow pass would have been able to control all the traffic in and out of the area. It should be restated here that this access into the Northern Sierras is one of the only few, and certainly the most direct, apart from the Coll de Soller route. So from the standpoint of trade this route would certainly have played a vital role in prehistoric times as well as historical ones so far as the mountain settlements are concerned. As will be shown when the finds from Matge are discussed, as well as recent finds in the Settlement Complex of Ferrandell-Oleza, this area was particularly rich both agriculturally and commercially.

The shelter is reached by way of a narrow path up the side of the mountain to the south of the S'Estret pass. This narrow path goes through a shrub-oak forest for about 3/4 of a kilometer before climbing a steep slope up the side of the mountain (**Puig de Boixes, 555m**) which leads to the rock shelter. At some unknown date the shelter was formed by a massive displacement of part of the north face of the Puig de Boixes, creating a natural overhang from about 10 meters to 30 meters high in some places (**see Plan 2A and 2B, Waldren 1982**). It seems that tectonic debris, probably cause by the original displacement and by subsequent erosion from the rock face, built up a natural platform in front of the newly formed shelter. As well be seen, this platform was later enlarged by the occupants of the shelter, who constructed a retaining wall along the edge of the platform, thus enlarging the area beneath the overhang and keeping the platform from eroding away downhill. As will be seen, this was done when the area was converted into a Talayotic Bronze Age cemetery. However, the earlier first settlers used the original platform as they found it, or else increased its width only rudimentally.

Undoubtedly, the site was initially selected by the ancient people who first settled there, circa 5000 b.c. to 4000 b.c., for its strategical situation within an attractive area as (**a**) it commanded a panoramic view of the valley of Valldemosa; (**b**) being closely situated to the narrow pass, it would have been quite easy to defend with only a small force of men; (**c**) the stream which runs through the pass would have been a convenient source of water during most if not all the year; (**d**) even today, the overhang gives ample protection from driving rain, despite the fact that further erosion of the rockface and other recent and more recent tectonic displacement has decreased the overhang in modern times; (**e**) at the same time, there is ample arable land available for agricultural and pastoral activities.

The different uses to which the shelter was put in prehistoric times as well as historical ones have been demonstrated during the excavations undertaken since it was first discovered by the author in 1968. There are five distinct purposes for which the shelter was employed; not necessarily continuous, between the limits indicated in any case:

(1) an animal shelter, including a corral for **Myotragus** between 5000 b.c. and 4000 b.c.

(2) a habitation for some of Mallorca's earliest settlers from about 5000 b.c. until circa 1400 b.c.

(3) a burial site during the local Bronze Age and Iron Age, circa 1400 b.c. to about 50 a.d.

(4) a workshop area during the local Beaker phase of the Pretalayotic Period, circa 2000 b.c. to 1400 b.c. and again during late Bronze Age of the Talayotic Period, circa 1000 b.c. to 8[illegible] b.c. as a pottery firing area.

(5) and finally, it was again used in historical and modern times as an animal shelter.

General Considerations

In the present chapter we are concerned only with the first and part of the second use of the shelter listed above, i.e. those related to the Presettlement and Early Settlement Period (**PSP and ESP respectively**). As has been noted earlier, the Matge stratigraphy is particularly strong in evidence concerning the Early Settlement Period. This evidence has not been the subject of any preceding section, consisting as it does of (**a**) mostly bones of ***Myotragus*** associated with evidence of early man at Matge and not at Muleta, (**b**) chronometric dating of these levels, (**c**) artefacts and traces of human activity not found at Muleta and (**d**), therefore, correlative as well as contemporary with levels at Muleta, and certainly alters the traditional view of Balearic prehistory and its chronological framework, especially when linked with the Early Settlement Period data from the Muleta deposit.

As will be shown, the ESP levels concerned in the Matge stratigraphy begin with the coprolite beds and ***Myotragus*** bones (**Plate 22:1, Valdren 1982**), circa 4730 b.c. and end circa 2700 b.c. These are the limits of the ESP, which evidently embraces all of the Atlantic stage of the Post Glacial Period, as it does at Muleta. It is this chronology as well as the specimen evidence supporting man's first occupation and activities in the Balearics that make it possible to correlate the two stations. After all, there is very little other than Munsell soil colour that the actual deposits at the two sites have in common. This is no doubt due to the fact that Muleta is a closed cave situation, while Matge is more or less an open-air site, so why should they be similar? They are some 20 kilometers apart. Then again, the exceptional depths that have been attained to date at Matge, although for the Early Settlement Period they are greater. Before the excavations can be taken any deeper at Matge, considerable engineering will be required to remove the overburden and make the underlying deposits accessible and safe for further excavation. Fortunately, this type of obstacle occurred only in the eastern end of the shelter and not in the other areas.

Actual naming of the different zones of the shelter's 46 meters by 10 meters area employed throughout its history did not occur until well after this important eastern part of the shelter reached an advanced stage of excavation, and it was impossible to advance the work any father in an easterly direction. At the same time, it was found that the length of the shelter could be divided into three main areas; and as it turned out, the use of the shelter by prehistoric man, in his different employment of the site, followed an eastern to western sequence and direction of utilisation. So, logically the areas of the shelter eventually took on names that were to later prove to be of help descriptively as well as in their separate study. The three areas of Matge are the following:

(1) **The Eastern Enclosure**, which measures an area covering about 20 meters by 10 meters, between survey markers 26-46 (**Plan 2A and 2B, Valdren 1982**). From the point of view of the evidence covering the use of this sector in prehistoric times, this is the most important area, and one which deals directly and exclusively with the Early Settlement Period, currently under discussion. It also plays an important role in the Pretalayotic Period, Talayotic and Post Talayotic Periods, having supplied most of the vital chronometric and artefact evidence.

(2) **The Central Enclosure**, which measures an area covering about 15 meters by 10 meters, between survey meter markers 11-26 (**Plan 2A and 2B, Valdren 1982**). The full importance of this area has not been understood until the present, when the study of its materials and further excavation has proved it to be a workshop area, circa 1800 b.c. to about 1400 b.c.

(3) The **Western Enclosure**, which measures an area covering about 11 meters by 8 meters, between survey meter markers 0-11 (**Plan 2A and 2B, Valdren 1982**). This is the last area to be used in the sector sequance, although toward the end of the shelter's use by prehistoric man there is evidence that he returned to using the Eastern Enclosure at about the time of the Roman Colonisation and afterwards for burial purposes.

Stratigraphical Contexts and Dating of the Matge Deposit: Summary

Although, for the most part, the methods of excavation at Matge can be considered conventional, the mature of some of the matrix in the various strata can be considered unusual and have necessitated different thinking and special treatment. One of these, that of inhumation in quicklime, has been already discussed in considerable detail in the section covering the methods used in dating (**Part I**). All the same, a brief description and a few words of discussion are required as an introduction to the various strata and the chronological periods they represent. For this purpose, the composition of the different strata and the compositional characteristics of the different prehistoric period into which they naturally fall can be outlined in the following four classifications, in reverse order of age. This will then give the reader a better idea of the varied stratigraphical conditions as they were encountered in the deposit. For reasons stated below the East Enclosure of the rock shelter best represents the various cultural periods encountered at Matge, and in so doing permits us to arrive at a good summary account of the stratigraphy.

The Post Talayotic Period, Iron Ages (Strata 6 to 1)

The East Enclosure contained the deepest and most complex deposit of lime conglomerate which in some sectors reached up to 175 centimeters in thickness, and never fell much below 100 centimeter deep (**Figures 4** and **5**). The lime conglomerate covered the entire East Enclosure of the shelter between meter markers 27 and 46., running from the inside of the base of the overhang wall right out to the Talayotic retaining wall as well as several boulders (**e.g. "J" rock and "P" rock**) which had fallen on top of the retaining wall in different sectors of the enclosure, after its completion by the Talayotic inhabitants and prior to the conversion of the East Enclosure to use as a quicklime inhumation cemetery. It is this East Enclosure which has given us the category study of the various conditions present in the vertical section of the lime conglomerate described in the author's thesis (**Valdren 1982**), and it is from these vertical sections that we have the best and clearest picture of the method itself.

In these profiles- too many to describe here- the various charcoal lenses separating the various burial strata can be seen to be composed of at least three different recycled master levels between meter markers 37 to 42, Stratum 6 to Stratum 2. It was in the East Enclosure that the first use as a Post Talayotic quicklime inhumation cemetery took place.

A summary of these Post Talayotic Iron Age stratigraphy is as follows:

(1) This is demonstrated by the area producing the oldest 14C date for the conglomerate of 2640 b.p. (**690 b.c.**) ± 100 years (**ABSM 23**)(**Stratum 6**), and represents the Early Iron Age in this sector. It is also the enclosure which produced most of the iron, bronze and other Post Talayotic artefacts; although it does not represent the oldest dated Early Iron Age contexts.

(2) The oldest Early Iron Age contexts originate from the dating of charcoal associated with iron artefacts is found beneath the first lime conglomerate level in the West Enclosure. In this area (**Meter Markers 1-11**) the dates signal at Matge the first and oldest introduction of iron. These two important dates are: Meter Marker 8; 2700 b.p. (**750 b.c.**) ± 80 years (**ABSM 25**) (**Stratum 8**) and Meter Marker 10: 2730 b.p. (**780 b.c.**) ±100 years (**ABSM 26**) (**Stratum 9**).

Other 14c dates for this level (**Stratum 6**) in other sectors of the enclosure are Meter Marker 37: 2540 b.p. (**590 b.c.**) ± 80 years (**ABSM 20**), Meter Marker 33: 2480 b.p. (**530 b.c.**) ± 70 years (**ABSM 17**).

In this sector Stratum 5 there was a thick charcoal lense in the conglomerate between 2 and 7 centimeters thick. Several dates are available associated with artefacts, depending on the sector. For example Meter Marker 22: 2520 b.p. (**570 b.c.**) ± 80 years (**ABSM 19**) of the Central Enclosure and Meter Marker 33: 2400 b.p. (**450 b.c.**) ± 80 years (**ABSM 14**).

(3) Stratum 4 in this sector is marker by a thick layer of lime conglomerate about 20 to 60 centimeters thick. It also contained abundant iron, bronze, lead and glass artefacts that were well preserved. This level can be generally dated by a bracket of dates on charcoal from other areas. For example: Meter Marker 8 of the West Enclosure, 2290 b.p. (**340 b.c.**) ±100 years (**ABSM 12**) ; Meter Marker 35 of the East Enclosure, 2240 b.p. (**290 b.c.**) ± 70 years (**ABSM 9**) and Meter Marker 36 of the East Enclosure 2200 b.p. (**250 b.c.**) ± 100 years (**ABSM 7**).

(4) The following level (**Stratum 3**) appeared in circular or pocket-like patches of charcoal in many areas of the East Enclosure as well as in the other enclosures. It is the charcoal of the last fires lit to reactivate areas of the lime conglomerate (**e.g. Meter Markers 39 to 42, Plan 2C: Valdren 1982**). It is this charcoal that has produced the youngest dates in the Post Talayotic contexts of this area of the cemetery. For example, two dates are the following: Meter Marker 31, 2080 b.p. (**130 b.c.**) ± 90 years (**ABSM 3**) and Meter Marker 22 in the Central Enclosure, 2070 b.p. (**120 b.c.**) ± 120 years (**ABSM 2**).

(5) The next two and last strata (**Stratum 2 and Stratum 1**) form the end of the sequence in the East Enclosure. Stratum 2 forms the pockets of lime conglomerate described above and consists usually of weathered lime residue if located deep under the overhang. They are pits dug into existing lime conglomerate and probably represent reactivated lime that has not been used, as no burials have been found in them. This also suggests that preparation of the quicklime may have been done periodically, as a matter of course. The ultimate stratum (**Stratum 1**) is a layer of modern-day debris and windblown earth accumulated over the area in places where the quicklime conglomerate contexts have been protected from the weather. In some areas for example between Meter Markers 36 to 44, a good quantity of classical pottery was found on the surface. The level is from 3 to 6 centimeters thick.

The Talayotic Period, Bronze Ages (Strata 10 to 7)

It is at this point in the rock sheleter' s history that the function of the Matge rock shelter changed radically from one of living in the Pretalayotic Period circa 1400 to circa 1300 b.c. into a cemetery in Talayotic times. It is also the point at present another alteration in function occurred as a radical change in burial practices; from the one we have just examined of inhumation in quicklime to cremation burials of the Talayotic Bronze Age. The exact cause for this abrupt change on either end of the interface of the Talayotic Period is not at all certain; although there is some indication seen in the stratigraphy by the position of many of the very large fallen pieces of the overhang, such as 'J' rock and 'U' rock etc., that the Pretalayotic inhabitants may well have considered the site unsuitable for living. In all events, the East Enclosure was abandoned completely for living some time after 1400 b.c.

The Matge Talayotic cemetery burial areas are quite extensive, being widely distributed over a large area of about 138 square meters in both the East and Central Enclosures, but being shallower in vertical profile, consisting of only at most 5 levels of about 50 centimeters depth. As can be seen in Plan 2 (**Waldren 1982, vol.III**), compared to the thick Post Talayotic quicklime burials which followed and covered the Talayotic cremations, these Talayotic levels are really very thin, sandwiched between the Pretalayotic habitation zones and the overlying Post Talayotic quicklime inhumations. It is this reason that the Talayotic stratigraphies can be more profitably studied in horizontal distribution than in a vertical one.

The necessity for studying the horizontal distribution of the Talayotic burials rather than their vertical stratigraphy has been brought about by several conditions. Initially, the burials appeared to have been laid out individually within the confines of the East Enclosure cemetery retaining walls, which were built to contain the cemetery. Excavation shows that originally these Talayotic burials were quite orderly, as can be seen in

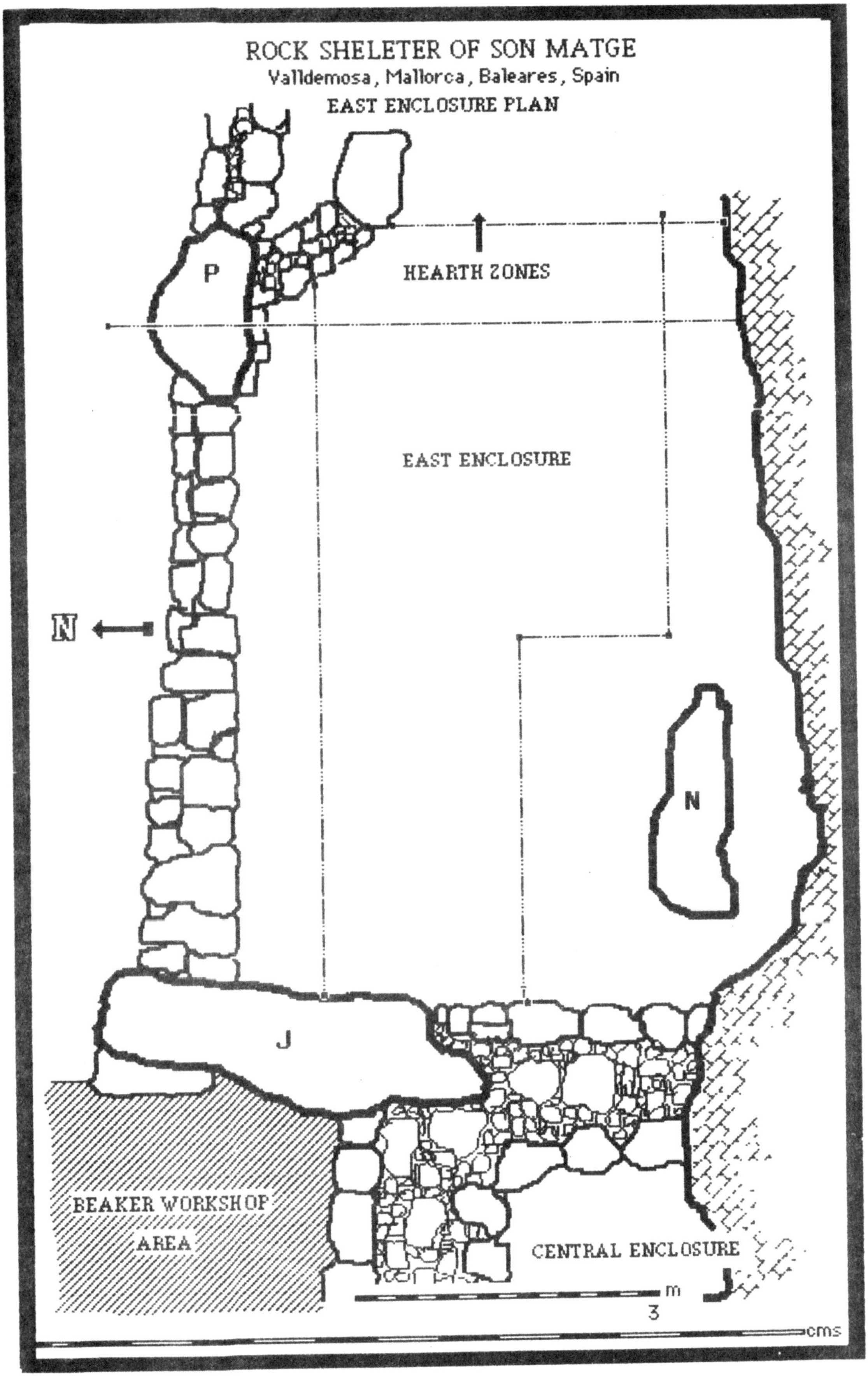
ROCK SHELETER OF SON MATGE
Valldemosa, Mallorca, Baleares, Spain
EAST ENCLOSURE PLAN
P
HEARTH ZONES
EAST ENCLOSURE
N
N
J
BEAKER WORKSHOP
AREA
CENTRAL ENCLOSURE
m
3
cms

FIGURE

4

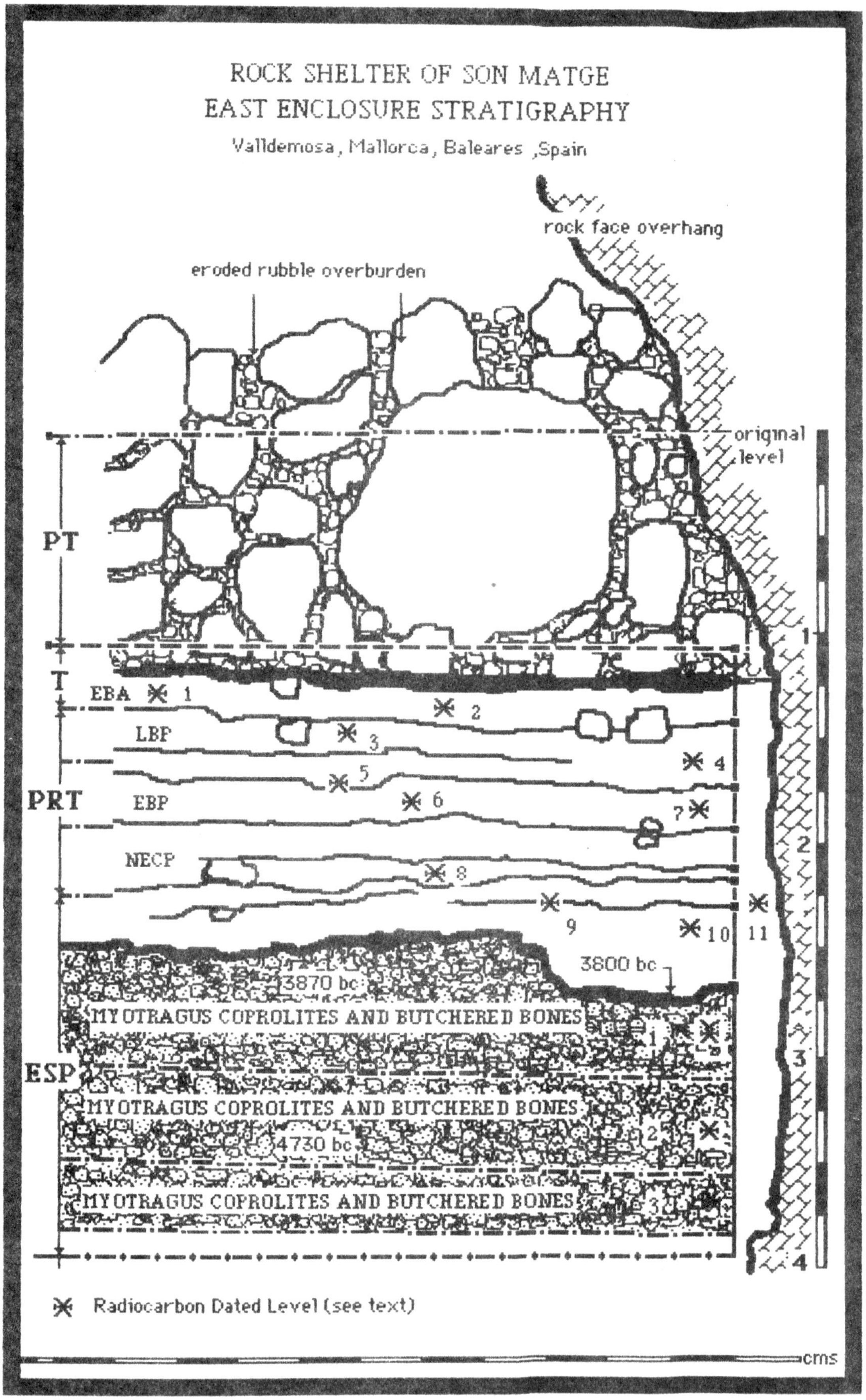
ROCK SHELTER OF SON MATGE
EAST ENCLOSURE STRATIGRAPHY
Valldemosa, Mallorca, Baleares, Spain
rock face overhang
eroded rubble overburden
original level
PT
T
EBA
LBP
PRT
EBP
NECP
1
2
3
4
5
6
7
8
9
10
11
3870 bc
3800 bc
MYOTRAGUS COPROLITES AND BUTCHERED BONES
ESP
MYOTRAGUS COPROLITES AND BUTCHERED BONES
4730 bc
MYOTRAGUS COPROLITES AND BUTCHERED BONES
Radiocarbon Dated Level (see text)
cms

FIGURE

5

the manner in which the pottery accompanying the dead was carefully placed in the burials. The individual burials were probably arranged in a linear manner, rather than covered by a shallow layer of earth. No actual grave appears to have been dug to contain the burial as the remains rested either on the shelter floor or directly on the Pretalayotic habitational levels. This accounts for the shallow stratigraphy throughout the entire shelter. The Post Talayotic overlying quicklime inhumations will be seen to have lacked this orderly arrangement of burial, and it appears to have been an essential characteristic of these Iron Age burials that the bodies were covered in quicklime which was periodically dug into in order to recycle and reactivate the lime after it was slaked, so that new burials could be made. This process is not infrequently distrubed and at times redistributed not merely the Post Talayotic burials but also the underlying Talayotic cremations. Therefore vertical stratigraphy does not exist or when it does it is quite complicated, and more can be gained from the study of the horizontal distributions of these two kinds of burial; although some consideration has been given to the vertical profiles, whenever they are important and good conditions exist, especially concerning the later Post Talayotic quicklime inhumations. It is the author's experience that similar straigraphical and physical conditions exist in most Talayotic and Post Talayotic cemeteries of this sort both in Mallorca and Menorca, and that failure to understand them has led in the past to misinterpretations of Talayotic and Post Talayotic burial customs. Accordingly, emphasis is placed on study of the horizontal distribution of the burials as well as the stratigraphical conditions, such as the Post Talayotic lime conglomerates, in each area with special attention on the horizontal chronological survey.

However before the pertinent levels are discussed, one further general point may be made regarding the primary research site of Matge as well as the secondary site of Muertos Gallard. At these sites not one bit of evidence exists to suggest that part of the offering accompanying the dead included any form of meat. On the basis of the otherwise accompaniments of abundant pottery in the form of 280 complete pots and reconstructable semi-complete and fragmentary ones in the order of many hundreds more, along with other grave goods, this lack of animal remains seems rather strange.

The stratigraphy of the East Enclosure is once more the most important of the three areas containing Talayotic evidence of the Bronze Age at Matge. It is also the best dated for this period as in the case of the Post Talayotic Iron Age. In fact as we will see this is also true for the dating and study of the Pretalayotic Period and the Early Settlement Period to be discussed presently. It also contained the best preserved artefact evidence, as indicated above, consisting of the enormous 'caches' of pottery. Undoubtedly, the angle of the overhang in this section of the shelter as well as the well constructed retaining wall which gave this enclosure its special preservation of materials, apart from the disturbance already mentioned caused by the Post Talayotic quicklime inhumations and other minor and major rock falls from the overhang.

The area involved in the Talayotic stratigraphical sequence during the Early, Middle and Late Bronze Age are defined by Meter Markers 27.5 to 44, areas east-west I1-I2 to K1-K2. In vertical section the stratigraphy covers a period of circa 1250 b.c. to 800 b.c.. A summary of the contents of these levels in the East Enclosure are the following: Four levels of various description which show burials dating from those strata stated above in which the method of cremation was practised. These cremation burials as compared to quicklime inhumations consist of burnt human bone and many bronze, but no iron items, and enormous quantities of pottery evidence of a <u>single</u> typological assemblage.

A Summary of the Talayotic stratigraphy is as follows:

(1) Stratum 10 is the lowest layer in this sequence and is made up of a mixture of very compact ash and earth and varying in thickness from 10 to 90 centimeters. It contained few artefacts and when it does these consist of an equal mixture of Pretalayotic common ware of the Late Beaker Phase or Talayotic wares of the Early and Middle Bronze Age and probably represent a contact interface between the two periods. The upper part of this level was very crusty in places as if it had been frequently trod upon, and may well represent a floor. In fact there are indications that this was the case; the Talayotic cemetery retaining wall was built directly on this layer, and activities involving the building of the cemetery retaining wall would certainly have resulted in the treading down of this ash and earth layer, left over from the last of the Pretalayotic kitchen fires in this sector of the East Enclosure. The level would have been equally compressed during the the actual process of preparing the Talayotic burials themselves. The area may have been first cleared or even ritually purified by making fires over the whole zone before the actual Talayotic burials took place. In all events, the level represents the first Talayotic stratum at Matge and also records the site's complete abandonment for living.

(2) Stratum 9 and Stratum 7 is a composite layer of charcoal, burnt and unburnt human bones from 10 to 20 centimeters deep. These levels appear at time to appear and diverge, separated by the red oxide burnt earth of Stratum 8. It is in these levels that charcoal associated with pottery have given us a date at Meter Marker 33: 3200 b.p. **(1250 b.c.)** ±100 years **(ABSM 36)**. This marks the only date for charcoal we have for the moment representing the Early and Middle Talayotic Bronze Age at Matge. The upper interface dating of the Late Bronze Age originates from the West Enclosure in the form of charcoal found in a pottery kiln in that area. This radiocarbon date is from Meter Marker 4: 2810 b.p. **(860 b.c.)** ± 50 years **(ABSM 28)**. Dating representing the interim between these two different interfaces of the Talayotic contexts at Matge come from other stations as we will see eventually.

The Pretalayotic Period, Copper Age, Neolithic (Strata 10 to 28)

The excavated area representing the Pretalayotic stratigraphy is quite extensive in the east Enclosure as well as in the other enclosure sectors. It covers an area of about 110 square meter and the depth of these Pretalayotic contexts are from 120 to 200 centimeters below the present-day ground level, depending on their location in the East Enclosure shelter area. The stratigraphy on this area is made up of an unbroken sequence of habitational hearth fire levels made up of successive layer of ash and charcoal extending down through the Pretalayotic Period and into that of the Early Settlement Period.

A summary of the contents of the Pretalayotic levels are as follows:

(1) Five levels of charcoal and ash representing hearth zones with frequent pottery evidence, domesticated animal bones along with ***Myotragus*** remains, all, showing signs of being burnt or butchered. These are the NECP levels of the Pretalayotic sequence and are made up of Strata 28, 27, 26, 25 and 24. Both the upper and lower interfaces of this phase have radiocarbon dating as well as Stratum 26. The dating sequence is Stratum 28: 4650 b.p. (**2700 b.c.**) ± 120 years (**ABSM 51**); Stratum 27: 4093 b.p. (**2143 b.c.**) ±392 years (**ABSM 50a**) and Stratum 26: 4020 b.p. (**2070 b.c.**) ± 50 years (**ABSM 50**).

(2) Seven levels of charcoal and ash representing hearth zones with frequent pottery evidence, domesticated animals but no ***Myotragus***. These are the EBP levels of the Pretalayotic sequence, containing abundant Beaker incised wares as well as undecorated Beaker wares and common wares. The sequence of the EBP is made up of Strata 23, 22, 21, 20, 19, 18, and 17. A chronometric series exists for this pahse. The dating sequence is as follows: Stratum 23: without 14C dating, Stratum 22: 3980 b.p. (**2030 b.c.**) ± 170 years (**ABSM 49**), Stratum 21: 3980 b.p. (**2020 b.c.**) ± 100 years (**ABSM 48**) Stratum 18: 3820 b.p. (**1870 b.c.**) ±120 years (**ABSM 46**) and Stratum 17: 3670 b.p. (**1750 b.c.**) ± 30 years (**ABSM 43a**).

(3) Eight levels of charcoal and ash representing hearth zones with frequent pottery finds, domesticated animals. These are the LBP contexts of the Pretalayotic Period containing abundant pottery fragments of the Late Beaker Phase and common wares. The LBP contexts are made up of Strata 16, 15, 14, 13, 12, 11, 10 and 9. A chronometric sequence exists for this phase. The dating sequence which is available for this series is as follows, Stratum 17: 3670 b.p. (**1720 b.c.**) ± 100 years (ABSM 43a), Stratum 16: 3620 b.p. (**1620 b.c.**) ± 80 years (**ABSM 42**), Stratum 14: 3480 b.p. (1620 b.c.) ± 80 years (**ABSM 41**), Stratum 11: 3380 b.p. (**1430 b.c.**) ± 50 years (**ABSM 38a**) Stratum 12: 3420 b.p. (**1470 b.c.**) ± 100 years (**ABSM 39**) and Stratum 11: 3350 b.p. (**1400 b.c.**) ± 60 years (**ABSM 38**).

The Early Settlement Period, Neolithic (Strata 36 to 26)

Until the present, the area excavated in the east Enclosure, in so much as the lower levels are concerned, lies at the extreme far end of the enclosure, consisting of an area of 35 square meters. The depth of these levels are about 250 to 400 meters from the original surface and represent the deepest and oldest levels so far reached at the site. The levels we are concerned with in the Early Settlement context are Strata 36 to 28, although the sequence actually ends with Straum 26: Stratum 26 being the interface or transitional level between the Early Settlement Period and the Early Neolithic Ceramic Phase of the Pretalayotic Period. A summary of the contents of these levels are as follows:

(1) Three levels of ***Myotragus balearicus*** coprolites with the animal's osteological remains, showing 'V' trimmed horn cores and numerous butchering marks caused by flint tools on the proximal end of long bones, where muscle and tendon attachments were located (**Plates 32:1 and 32:2, Waldren 1982**). Strata 36, 35 and 34 represent these contexts.

While Stratum 36 is the lowest and oldest of the levels so far reached in the enclosure, the animal remains that have been found have as yet not been dated. The soil, like the two superior ones, is made up of a fine clay-like dry soil mixed with rather smooth oval shaped stones, which appear by some agency to have been weathered, ranging in size from 5 to 6 centimeters in diameter. The soil part of the matrix is made up of a large quantitiy of coprolites, both complete and fragmented and both of which have a colour of **Munsell** 7.5YR, 5/4, dry. The stratum is about 40 centimeters thick, with its top about 390 centimeters from the shelters original level in this part of the enclosure. The fact that the coprolites were well preserved in shape- although they were and are dry and fragile today- is die to the presence of the stones; the coprolites having slipped down and collected in the air-spaces between the stones. As stated above the exact nature of the stones of the layer and the subsequent ones is not known. They do have a slight coating of calcium carbonate and it may be the agency of percolating water that is responsible for the shaping of these stones. In some ways they are like '**scree**' material because of their uniform size and may well be the end product of the original deterioration of the walls of the shelter overhang. The stratum also contained butchered bones of ***Myotragus*** and a few '**V**' trimmed horn cores.

Stratum 35 is the second of these three coprolite beds distinguishing itself from the inferior one only in the matter of soil colour, **Munsell** 10YR, 7/4, dry, and the fact that one radiocarbon date on bones of the ***Myotragus*** with butchering marks and 'V' trimmed horns for this level does exist. This 14 C date is Stratum 35: 6680 b.p. (**4730 b.c.**) ±120 years (**ABSM 55**). The level is like the former one about 40 centimeters thick.

Stratum 34 is the third of the series of three coprolite beds with a similar matrix as the preceeding ones and starts at 310 centimeters below the surface of the shelter. The soil colour is again different that the other two, Munsell, 7YR, 6/4. Coproilites from the upper part of this level have been dated at Stratum 34: 5820 b.p. (**3870 b.c.**) ± 360 years (**ABSM 53**).

The reader should bear in mind that as of the present no post holes or other evidence of a corral structure delineating the extent or form it took. This is believed to be due to the quantity and nature of the debris that still must be removed, as well as the fact that the Talayotic cemetery retaining wall still run over the majority of the corral area. These features must be removed before the full extent of the corral and other superior living areas can be determined. In fact, there may never have been a permanent structure to this corral; that is to say post having been erected in the soils. A corral structure could have been erected by piling brush or even stone walls- as much stone debris is on hand even today and such structure would not necessarily have survived.

However on the basis of the coprolite beds and soil colour changes as well as the radiocarbon documentation, it would appear that the shelter's function as a corral for ***Myotragus*** lasted some 900 years.

(2) The rest of the Early Settlement Period stratigraphy is made up of six levels of compacted ash and charcoal; occupational levels in the form of the first of the hearth zones at Matge. These contained butchered ***Myotragus*** bones, the youngest of which had pottery evidence and kitchen debris of domesticated animals (**sheep, goat, pig and cattle**). However as this is the upper interface between two chronological periods, only those that are '**aceramic**' and without domesticated animal evidence are discussed in this ESP sequence.

Stratum 33 is the first of the hearth zones and marks the first use of the shelter as a habitation zone. The level contained the first burnt ***Myotragus*** bones (**Plates 32:1 and 32:2, Valdren 1982**). The level ran along the length of the bottom of the excavated zone between Meter Markers 42.5 to 46. Thus far the level has contained no artefacts, apart from one flint nucleus, found at Meter Marker 44. A radiocarbon date on charcoal from this bottom-most hearth gave a reading, Stratum 33: 5750 b.p. (**3800 b.c.**) ±115 years (**ABSM 43**).

It is interesting to recall here the 14C age of the earliest human remains at Muleta, 5935 b.p. (**3985 b.c.**) ± 109 (**SM 54**) and further note that the central figures fall no more than 185 years apart; thus establishing a correlation within the period.

Stratum 32 consisted of densely compressed ash from a long succession of fires of domestic origin, about 10 centimenters thick. There are several arrangement of hearth stones that contained only burn ***Myotragus balearicus*** bone. These hearth stone arrangements (**Figure 26, Hearth 9, Valdren 1982**) are the first clearly defined hearths. As they appear directly over Stratum 33 which contained the charcoal of Stratum 32, they could actually include the first of the kindled fires of Hearth 9 here.

Stratum 31 through 32 are a continuance of those of the Early Settlement sequence. The desciption of these are a repeated condition of compressed ash and charcoal levels and ***Myotragus*** kitchen debris continuing to Stratum 28.

(3) These levels of compressed ash and charcoal continue from Stratum 28 to Stratum 26 of the Pretalayotic Period. Important here is Stratum 28 where a significant radiocarbon date is available for the introduction of pottery and domesticated animals into the stratigraphical contexts. This date for Stratum 28: 4650 b.p. (**2700 b.c.**) ± 120 years (**ABSM 51**) dates the kitchen fires of Hearth 5.

Stratum 26 deserves some inclusion in the form of a few remarks here in regard to the final evidence of the ***Myotragus*** in the stratigraphical and chronological sequence at Matge, despite the fact that it marks the upper interface of the NECP of the Pretalayotic. For it is this level which suggests the final extinction date of the animal much later than hitherto expected (**Burleigh and Clutton Brock 1980**) and dated at 4093 b.p. (**2143 b.c.**) ± 360 years (**ABSM 50a**). At the same time bring to a close the question of its inclusion in this section.

These latter stratigraphical contexts of the Early Settlement Period contexts at Matge and also signal the close of the discussion of the primary sites as relaled to this station. It relates, too, the stage at which the investigation of the Muleta rock shelter remains at present. There is still work to be done at Matge on these lower levels when the dangers of heavy depositional over-burden are removed and work can once more begin. Until the time that further work is underway, the current summary brings the findings up to date for this station and its various stratigraphies. Some further discussion of the different contexts are found in the observations and conclusions subsequently presented (**Part IV**).

The Ferrandell-Oleza Settlement Complex (1, C, SFO-OS and SFO-YS) (6°.16'E; 39°.43'N)

Introduction

<<<<<<<<<<>>>>>>>>>>

As we have seen from the evidence and other data discussed above, the majority of the information in terms of chronometric dating and artefact evidence has come from cave and rock shelter settlements. However, it is the larger more complex establishments of Pretalayotic open-air communities from which we can expect a clearer perspective as to early organised Balearic society and the economic conditions present and at work in such sites.

The aspect of Pretalayotic Copper Age open-air settlement in the Balearic Islands existing prior to the settlement of the Talayotic Bronze Age is in many respects a completely new one. This lack of Pretalayotic open-air settlement in the past has led some scholars of Balearic prehistory to estimate a high antiquity for the Talayotic settlements, (**Pericot 1973**); placing the age of some of their buildings, including the well know structures and reputedly oldest called navetas to circa 1800 b.c. to circa 2000 b.c. This as indicated from recent surveys, in which radiocarbon dating has played an important role, is just not the case. As we will see, Balearic open-air settlement is a Pretalayotic Chalcolithic event which at least in one instance can be demonstrated as having been present as early as 2000 b.c., indicating that the still higher organised Talayotic communities of the Bronze Age should be assessed as being closer in age to circa 1300 b.c. to circa 1200 b.c.

The main reason for the hiatus between what we have come to know as the '**parent**' Pretalayotic Culture of the Caves and the development of the full grown Talayotic Bronze Age culture has been simply a case of insufficient evidence which could provide us with the necessary missing details. Consequently, investigators up until now, when they considered any Copper Age contexts at all, or for that matter contemplated the origin of |the Talayotic culture itself, have had to draw heavily on the incomplete evidence of the graves and small cave and rock shelter living areas for insight into such subjects. Needless to say, such a hiatus has hindered proper interpretation and caused no end of difficulty in understanding the development of events in the islands as they probably occurred. However, over the last decade, since the discovery of sites like Muleta and Matge, details regarding early open-air settlement as well as that of cave dwelling communities have begun gradually to reveal themselves; especially with the advent of long-term surveys linked with radiocarbon dating.

The first documented indication of Pretalayotic open-air settlement emerged when G. Rossello Bordoy (**Cantarellas Camps 1972 and Rossello Bordoy 1973**) reported on the Bell Beaker sherd evidence from an open-air situation at Ca Na Cotxera in which radiocarbon analysis of an associated level of charcoal just above a pottery bearing one gave a date of 3750 b.p. (**1800 b.c.**)± 120 bc. (**CX 44**). Although no Pretalayotic architectural features were discernable to account for the Beaker contexts, there were adjacent building elements evaluated as Talayotic in origin. Unfortunately, further pursuit of the extent of these early Pretalayotic open-air contexts was not undertaken.

The Geographic Situation of the Ferrandell-Oleza Settlement Complex

<<<<<<<<<<<<<<<<<<<<<<<<<<<<<<<<<<<<<<<<<<<<<<<<<<<<<<<<<<<<<<>>>

Situated in the Jurassic Northern Sierras of Mallorca, near the Village of Valldemosa, at the foot of a limestone ridge that overlooks the Mediterranean and at the extreme western end of an alluvial plain, known locally as 'El Pla del Rey' of 'Plain of the King', the **Ferrandell-Oleza Copper Age Old Settlement and the Bronze and Iron Age Younger Settlement** forms part of what consists of a much larger prehistoric '**settlement complex**'; one which extends in this sector of the wide alluvial plain for nearly 1000 meters in a general north to south direction (**Figure 11**).

An even large overall settlement pattern of the 'Pla del Rey' is expressed by the extensive prehistoric architectural evidence and long term exploitation of the rich alluvial soils of the plain throughout most of Balearic Prehistory. Although the aspects of long term exploitation of the plain are of particular significance and concern in the conclusions arrived at later in this paper, they are only touched on briefly here in order to set the geographical and environmental stage for the reader. And it must be borne in mind that while in ground survey the entire circumference of the 'Pla del Rey' shows very extensive prehistoric occupation in every direction, this present report deals only with the excavation and radiocarbon dating documention of only a very small geographic zone; that of the **Ferrandell-Oleza Settlement Complex** at the western end of the plain.

Concerning the General Stratigraphy and Architectural Features

Typical of most Western Mediterranean early open-air settlements, but particualrly so with those of the Copper and Early Bronze Age of Southern France and the peripheral areas of the Spanish Mainland, the stratigraphy of such sites as the Ferrandell-Oleza Settlement Complex is generally one of a highly eroded nature. Vertical stratigraphy is limited to a depth seldom exceeding 100 centimeters and in most case as shallow as 25-30 centimeters in which no more than four or five levels are discernable in open areas. It is this eroded nature of these early settlements on the whole, as well as the fact that in most cases such stratigraphies are further complicated by their having undergone cultural intermixing by successive habitation and agricultural activity, that leads to complicated problems of interpretation. However, despite the fact that almost without exception the fields and other elevated areas of the SFO-OS have been subjected to a great deal of natural erosion, as well as heavy agricultural and other activities over the millennia, the site has not undergone the superimposition of successive levels. And while in most cases the upper-most rows of ancient stones of the structures have been destroyed for their building material potential in the construction of modern agricultural terraces, as well as having been used to make gravel for road building as late as 1956, the destruction has been quite straight forward and not complicated by context intermixing. In short, a good proportion of the architectural foundation stones are still in place. In other instances, field stone loosened by ploughing has been accumulated over many hundreds of years into what are locally known as '**clapers**', or the local name for stone piles, which have been stacked by farmers over the ancient structures; thus obscuring them, but equally preserving them.

This latter 'claper' accumulation has resulted in many of the original foundations of the ancient walls and other constructions being reasonably well preserved, having limited ploughing and other field activities over many centuries. At the same time, very little actual soil over-burden accumulation has occurred making heavy excavation necessary to arrive at the building foundation stones and other archaeological levels. It is only a matter of removing the piled stones one by one in order to reach the archaeological layers or carefully remove what soil over-burden that does exist.

In summary, while the vertical stratigraphy as found in the Ferrandell-Oleza Settlement Complex is particularly poor in depth, especially in regard to the oldest contexts, there is certain compensation in the fact that the vertical stratigraphy which does exist is fortunately not in the least complicated in the normal sense; consisting of thin levels of varying thickness and complexity. Another factor, established early in the excavations, is that the vertical stratigraphy also can be counted on usually to produce materials belonging only to a single specific horizon; thus in a sense also facilitating the interpretation of any given area. Admittedly, still on the other hand, the delicately '**weathered in**' nature and thin '**stratigraphic lenses**' that result in such eroded sites can be equally difficult to interpret, or even more difficult in other aspects, than those of broader and thicker stratigraphical accumulation. However, the author finds some advantages and preferences in the shallower vertical stratigraphy when it comes to horizontal stratigraphic distribution and interpretation of the material throughout; as well as being able to arrive with less effort at the pertinent archaeological horizons.

Remarks on Methodology, Preservation, Artefacts and Distribution

In fact, the aspect of the state of stratigraphical along with the accompanying architectural preservation in early Pretalayotic open-air settlements is a major point of consideration in any future prospection, excavation and subsequent study of these open-air communities. Experience in the Pretalayotic Settlement Complex of Ferrandell-Oleza shows that the architectural remains of such local, early open-air settlements can be expected to be very badly preserved; having at most what the author calls a '**low profile**' stratigraphy. This is to say, stratigraphical contexts and architectural remains consisting of only one or two distinct archaeological levels and courses of foundation stones with a stratigraphy that is for the most part horizontal rather than vertical, and most certainly severely eroded.

The author's experience at the Ferrandell-Oleza Pretalayotic settlement over the last seven years has shown that the site's development pattern through time occurred in a '**linear**' manner, hence, defining it as a '**linear settlement**'; one in which occupation has taken place in different geographical areas at different chronological times (**Waldren 1980**). This factor of linear settlement pattern along with '**low profile stratigraphy**' may well account for the fact that until the present so few Pretalayotic open-air settlements have been found.

One of the most important features of the Ferrandell-Oleza Settlement Complex has been this linear pattern of development, where the Pretalayotic Old Settlement (**SFO-OS**) area is quite independent and separated from the later Talayotic Younger Settlement (**SFO-YS**) zones. A factor which, more than any other single one, has helped not only, the assessments of the site's settlement pattern as a whole but, also, in the interpretation of the chronological periods involved. To exemplify this '**linear**' pattern of development at the site, nearly two hundred meters geographically and several hundred years chronologically separate the site's Pretalayotic Old Settlement from its Talayotic Younger Settlement.

At the same time, there are other advantages, as briefly pointed out earlier, in excavating such widely separated geographic areas in that each of the settlement zones have their individual and particular architecture and artefacts; thus greatly simplifying their systematic study, especially in terms of distribution, typology and chronology. These relatively uncomplicated conditions make it easier to assess specific disciplines, despite the lack of normal vertical stratigraphy. For example, they foster the use of computer analyses in the study of the horizontal distribution of artefacts and their attributes in the various areas and methodological approaches in lieu of establishing information as to the site's economic and social organizsation, as well as technological developments during the different prehistoric periods on the site. (**e.g. Ensenat and Waldren 1986 and Morell Orlandis and Querol Sagasta 1986**).

Although similar stratigraphical and distributional conditions are so far of rare occurrence, and may not occur in quite the same way again elsewhere, such '**low profile**' and '**linear**' stratigraphical characteristics in the prospection of Pretalayotic open-air settlement should be given serious consideration in the future search of potential communities of similar age. Experience in the Ferrandell-Oleza Settlement Complex has demonstrated on the basis of recent findings that consideration should be given to geographic areas outside the perimeters of the later Talayotic settlements themselves. This experience has also shown that special attention in methods of prospection and excavation should also be undertaken where the investigations and site excavational strategy are planned around a number of small exploratory trenches excavated in areas away from the more obvious and better preserved younger Talayotic architectural features, before any full scale operation is initiated; a method which, over the first few years in the Ferrandell-Oleza Settlement Complex, has been very successfully applied.

Another important factor, also a lesson learned from the initial survey trenches in the Ferrandell-Oleza settlement, is the apparent uniform distribution of the Copper Age occupational debris over a wide area of the Old Settlement site. In stratigraphical terms, the vertical distribution, which seldom exceeds 100 centimeters, and for the overwhelming part is about 25 centimeters in depth, is about the same all over the site complex, regardless of the general chronological context. It is then the dichotomous nature of these conditions which are advantageous on the one hand in making some interpretation easy and on the other hand equally more difficult. For special consideration has had to be put into operation in compensation for the lack of the usually expected more conventional vertical stratigraphy.

This special consideration then has not only centred on the delicate interfaces between the layers throughout the site, which under such eroded circumstances never are as easily discernable as would be wished, but has also centred on the spatial distribution of materials and other evidence in a computer based recording method, carried out on tightly controlled square meter by square meter excavation (**Waldren and Ensenat 1986**). A recording method in which the attributes of each excavated meter, its artefacts, animal bone and other information have been plotted and computer recorded as to distribution, function and typology. Other approaches also are being carried out, employing methods such as pottery thin sectioning, chemical analysis of the clays and duplication experiments (**Waldren 1982, Appendix 3A**).

The excavational strategy of prospection in a widely spread number of areas throughout the settlement complex has shown that the pottery evidence, especially, could be relied upon in the early survey excavations to be more or less distributed near to and related with the foundation stones of the architectural features throughout the different occupational areas of the site; a natural enough expectation as the buried architectural elements can be counted on to trap artefacts, like pottery and bone eroding downhill. All the same, this has been helpful, not only, in quickly assessing the overall duration of occupation of the site, as one spanning most of the Balearic prehistoric periods, by dating organic and other materials associated with the individual zones or larger areas and comparing the information with that from others, it has been also an aide in the establishing of the various stages of architectural construction that took place on the site; therefore, demonstrating reasonably early in the excavations not only the site's long period of occupation, but also revealing its complexity.

From the standpoint of material evidence in the Ferrandell-Oleza Old Settlement, the most outstanding sample lies in the large quantities of fine and common wares of the same typology as those usually found in the Pretalayotic context of caves and rock shelters as well as the recently associated with megalithic tombs (**Fernandez and Topp 1984**). These consist of pottery with horizontally and vertically perforated lugs and lug-handles as pottery features and incised or carved relief or corded borders of the finer wares ; diagnostic of Pretalayotic pottery diagnostic as early as circa 3000 b.c. to circa 2000 b.c. and which survived as late as circa 1400 b.c.

However, especially unique is the remarkably large amount and variety of finely incised Beaker wares found widely distributed in all the excavated areas of the site; a factor, which challenges the '**exclusiveness**' of this pottery and suggests a much more common use of these hitherto polemic wares. The site has produced hundreds of incised Beaker sherds; many of which resemble very early types (**Valdren 1984 and 1986**). Accompanying the Beaker pottery finds are also other items generally found in the Bell Beaker Assemblage; for example the **V** perforated bone buttons, **archers** wristguards, copper awls and tabular flint sickles and blades. An artefact assemblage for which the radiocarbon surveys have done much in assessing.

Along with the rich pottery and other artefact evidence, the excavated areas of the site have also provided us with abundant domesticated animal remains of goat, sheep, pig, small cattle and dog, all of which bear testimony to an active, productive and well rounded pastoral and agricultural economy; one which appears very different from the sparse and relatively simplified social and economic organisation suggested on the basis of the evidence of the caves, rock shelters and tombs.

Overall , the frequency and apparent common circulation and possession of the Balearic Beaker pottery and other artefacts, generally considered both **prestigious** and **exclusive**, the impression is one of a quite sophisticated, solid economic structure; one in which surplus was created and therefore present at least in specific levels of the society. The rich and capital invested architectural of the Ferrandell-Oleza Old Settlement further support and foster an **exclusivity**. The question of **elites**, the result of social divisions as currently argued by some investigators (**Harrison 1981**), may actually have occurred as the evidence of the site suggests.

However, as a counter- argument to the one of 'elites' in which only the wealthy could afford such rich goods, there is the disconcerting factor of the **common** possession of such prestige objects among the general population of the Pretalayotic rock shelters and caves settlements. In short we are not sure whether or not the Ferrandell-Oleza Old Settlement was typical of those that were in general existence at the time. As mentioned above, the assessment of the exclusivity of such items as rich pottery, becomes even more problematic when one does consider the obvious spread of these **rich** commodities into such far reaching areas as caves and rock shelters occupied by small family groups. But then again this may actually have been the state of the overall social and economic structure of the times, in which such items were much more easily accessible than generally asssumed.

Whatever the case, the fact remains that based on the physical evidence, the standard of living in the Ferrandell-Oleza Old Settlement was indeed high, no matter how one looks at it. The very fact that what evidence we do have is a very fragmentary proportion of the actual items that may have been available; emphasising the idea of wealth even further.

However, the strongest argument for '**elite**" or '**wealth**' in a small open-air community, certainly, finds reinforcement in the architectural construction of such an extensive and elaborate open-air settlement as that of the older areas of Ferrandell-Oleza. As we shall see below, the site with its two-meter thick, large stone wall enclosure and equally well constructed inner habitations and water system, built over an area of over 3600 square meters, not including extensive outer agricultural areas, strongly indicates the capabilities of a well organised community. A well organised community which succeeded in making its presence known; as well as establishing its particular exclusivity. In all events such a settlement represents a permanent investment and declaration of '**proprietorial domaine**' and indicates strong social and economic divisions within the populace. A single establishment which in this case has been delineated by a series of on-going radiocarbon surveys during an occupational duration of some 700 years (**Table 16**).

The Ferrandell Oleza Settlement Complex Architectural Elements and Their Related Dating

<<<<<<<<<<<<<<<<<<<<<<<<<>>>>>>>>>>>>>>>>>>>>>>>>>

The Ferrandell-Oleza Settlement Complex with its Older and Younger Settlements is spread out over approximately 1000 meters at the foot of an escarpment at the west end of the fertile 'Pla del Rey', Plain of the King. It is made up of an extensive series of what were once well built walls, a wide variety of buildings and other constructions which are examined below. The two settlement areas are examined in order of age beginning with the Pretalayotic Old Settlement. The architectural features from each of these settlements are discribed in much more detail in recent reports. Further description concerning these can be obtained from those publications (**Waldren 1984 and 1985**). However, some essential general details are found in the individual summaries below, as they seem pertinent. A summary of the pertinent geographic and stratigraphical contexts are as follows

The Pretalayotic Old Settlement

‹‹‹‹‹‹‹‹‹‹‹‹‹‹‹‹‹‹‹‹‹‹‹‹‹‹‹‹‹‹››››››››››››››››››››››››››››››

The Pretalayotic (**Copper Age**) Old Settlement consists of an enclosed large stone wall compound, rectangular in shaped with over two meter wide foundations that rose originally to an estimated height of over three meters. The four walls delineate a walled-in area of 3606 square meters.

The Compound Inner Areas can be divided into two main areas (**1**) the Domestic Zones, consisting of two or possibly three stone dwellings Apsidal Habitation 4 (**H4**), the Central Stone Structure (**CSS**) and one recently found poorly constructed semicircular dwelling in the western zone. There is a elevated Water Channel (**WC**) made up of aligned large stone slabs, at present excavation over 40 meters long, which traverses and divides the western half of the compound. The Water Channel has a downhill catch basin and an uphill stone walled reservoir attached to the northwest corner of the West Compound Wall (**Figures 6 and 7**) and (**2**) the Work Areas, a large flat area in the eastern half of the compound with signs of small dwellings or lean-to structures built against the north and south ends on the inside of the East Compound Wall.

The Old Settlement's interior structures with its various major architectural features are associated with radiocarbon dates analysed from charcoal and domesticated animal bone samples. On the basis of radiocarbon analyses in these areas, a duration of occupation of over 700 years is indicated.

The stratigraphical contexts and other related data is as follows:

(**1**) The **South Compound Wall** is 43.50 meters long and contains the main entrance into the Compound proper. This **South Compound Wall Entrance** is incorporated into the South Compound Wall as a tower-like construction. The **Tower Structure** itself is an enlargement in the centre sector of the South Compound Wall and consists of a tunnel like entrance with a small inner chamber, thought to be quard chamber for a man or dog. A radiocarbon date on animal bone from the superior most level to the northwest of this structure and believed to be an abandonment date has given us the following reading: Stratum 1: 3095 b.p. **(1145 b.c.)** ± 70 years **(SFO-OS 35a)**.

(2) The **East Compound Wall** is 69.5 meters long, running almost due north. It is along this East Compound Wall that charcoal from a hearth area at the base of its inner foundation gave a date of Stratum 3: 3950 b.p. **(2000 b.c.)** ± 65 years **(SFO-OS 47a)**; a reading that possibly represent a wall construction date in this quarter. On the basis of heavy pottery distribution along the wall, there appears to have been extensive occupation activity along the inside foundation, both, in the northern end as well as the outside foundation at the south end of the wall . It is suggested that this may have been the refuse from earlier living areas or lean-to structures contemporary with the wall itself.

(3) The **North Compound Wall** is 49.5 meters long. This compound wall also has an entrance **(North Compound wall Entrance)** of simpler construction half way along its length. Although wider than the South Compound Wall Entrance, it probably was used as an access for animals into the eastern areas of the enclosure. A radiocarbon date for charcoal from a fire at the base of the west side of the North Compound Wall Entrance, also perhaps of construction age, has given us a date for Stratum 3: 3790 b.p. **(1840 b.c.)** ± 90 years **(SFO-OS 45a)**.

(4) The **West Compound Wall** is 72 meters long. This wall is for the most part unexcavated, although its full extension has been established. No dates are currently available, although samples have been collected for future analyses from similar circumstances as those of the other walls.

(5) The oldest of the interior structure dated apart from the compound walls themselves, probably representing an occupation level or constructional fill, is from an ancient terrace **(Ancient Terrace 1-2)**. Here a date for Stratum 2: 3480 b.p. **(1530 b.c.)**± 80 years **(SFO-OS 41)** is available.

(6) Debris from the inside of the elevated **Water Channel**, which divides and traverses the site's occupational areas and runs alongside the **Central Stone Structure**, has given us another date for the site's abandonment; more than likely debris from the ultimate occupational period. This date for the fill debris of the Water Channel, consisting of domesticated animal bone, is Stratum 2: 3210 b.p. **(1260 b.c.)** ± 80 years **(SFO-OS 36a)**. Still another date also representing abandonment analysed on domesticated animal bone from a surface level comes from the southwestern end of the Central Stone Structure in Stratum 1: 3070 b.p. **(1120 b.c.)** ±50 years **(SFO-OS 34c)**.

7) Several other structures within the compound, which served as living and work areas, have also been excavated and organic remains in the form of charcoal and animal bones collected from beneath their foundations for analyses. These important dates are pending, but should add interesting data to the entire sequence for the Inner Compound when completed. As the test samples were collected from beneath these structures, they should give us readings which delineate the structures prior to their construction.

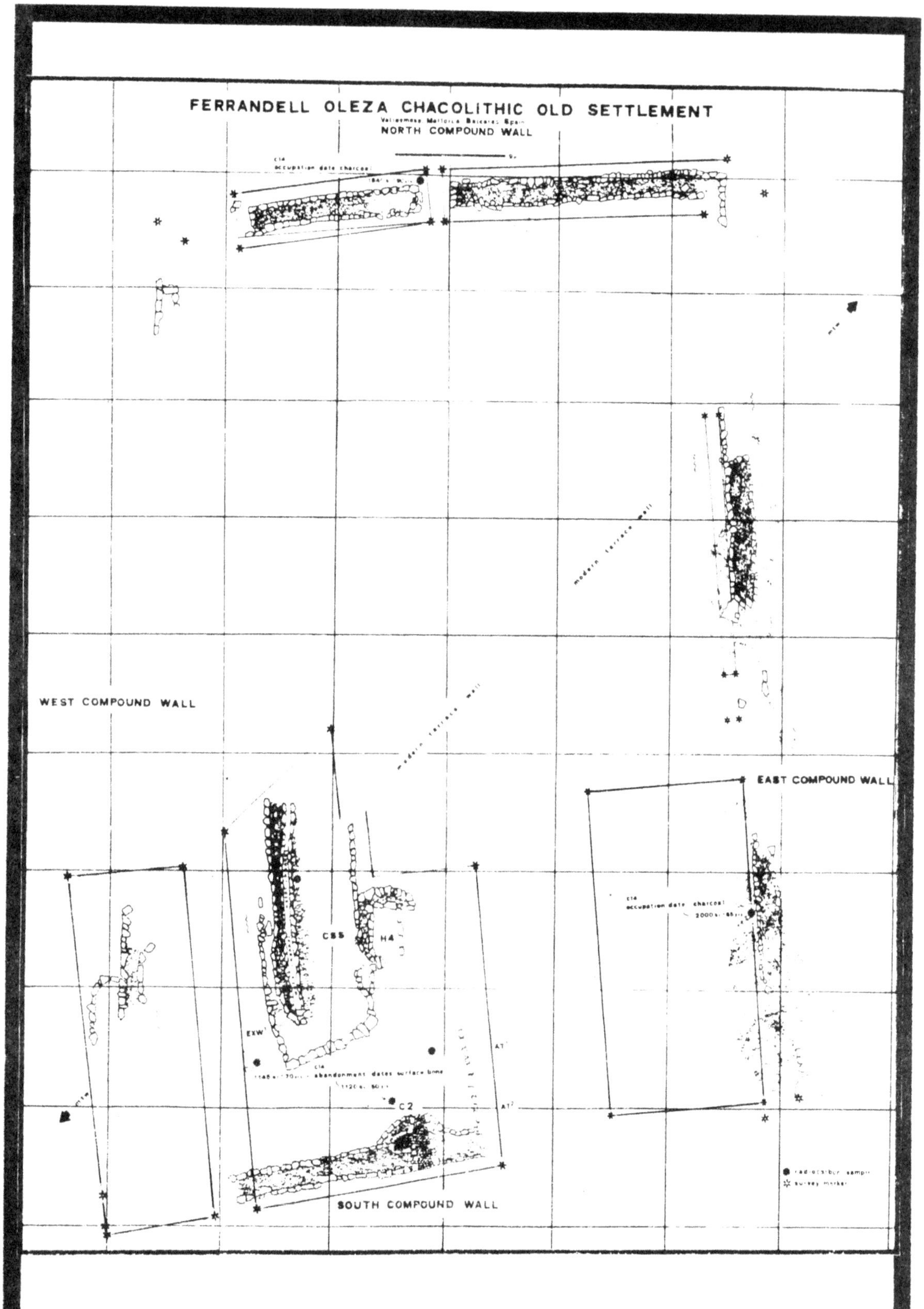

FERRANDELL OLEZA CHACOLITHIC OLD SETTLEMENT
NORTH COMPOUND WALL
WEST COMPOUND WALL
EAST COMPOUND WALL
SOUTH COMPOUND WALL
CSS
H4
C2
EXW

FIGURE

6

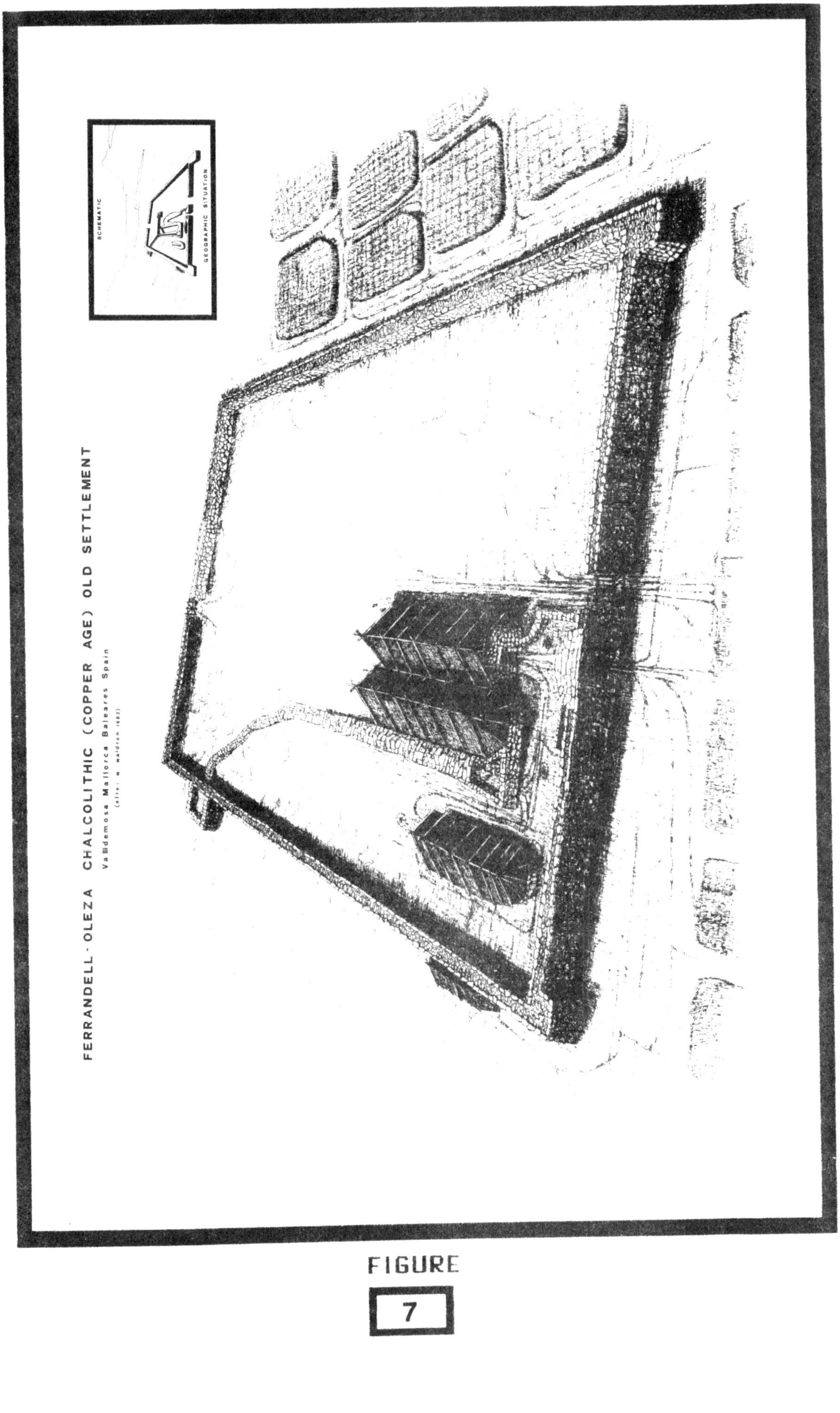
FERRANDELL - OLEZA CHALCOLITHIC (COPPER AGE) OLD SETTLEMENT
Valldemosa Mallorca Baleares Spain
SCHEMATIC
GEOGRAPHIC SITUATION

FIGURE

7

The West Old Settlement

In a sector some 60 meters to the southwest of the precincts of the Old Settlement Compound , an area of approximately 250 square meters has been excavated (**West Old Settlement**) and a special computer distributional study made (**Ensenat and Waldren 1986**). This area which on the basis of heavily accumulated debris, with no discernable architectural features, but made up of abundant remains of Copper Age pottery and other artefacts, is believed to be a refuse area. Such areas are known and have been reported on in the past as excellent artefact study areas (**Shennan 1974**). The excavations in this area has given us two very closely correlated radiocarbon dates of important antiquity as well definition to the Copper Age contexts of the Ferrandell-Olexa Settlement Complex in the form of artefacts, which are identical to those found within the compound walls of the Old Settlement.

(**7**) Here, in the West Old Settlement these two charcoal dates in both cases are for two different zones in Zone 1 Stratum 3: 3720 b.p. (**1770 b.c.**) ± 35 years (**SFO-OS 43b**) and Zone 2 Stratum 3: 3700 b.p. (**1750 b.c.**) ± 30 years (**SFO-OS 43c**). These radiocarbon dates are associated with various artefacts of Beaker assemblage and typology.

(**8**) There are a further series of dates also all from surface materials, the result of relatively modern origin from their location and age ; activities such as modern terrace wall construction and field clearing work. These dates are: 1650 b.p. (**300 a.d.**) ± 80 years (**SFO-OS 1b**), 1670 b.p. (**280 a.d.**) ± 15 years (**SFO-OS 1c**) for charcoal from field fires. Another group is: 1710 b.p. (240 a.d.) ± 60 years (**SFO-OS 1d**) and 1730 b.p. (**220 a.d.**) ± 12 years (**SFO-OS 1e**) from charcoal at the base of a modern terrace wall, again the result of modern field work activities.

The pertinent stratigraphical and architectural contexts along with the chronometric data summarised above represents the current stage of excavations in the Ferrandell-Oleza Old Settlement.

The Talayotic and Post Talayotic Younger Settlements

The Younger Settlement areas are located about 100 meters to the south of the Old Settlement Compound, and consist of five of the megalithic stone structures known as Talayots and diverse adjoining architecture. All the structures in the Younger Settlement excavated todate are of Bronze and Iron Age origin. While most of the emphasis until now has been centered on the excavation of the Copper Age Old Settlement, some work has been carried out in a selected number of locales within the Younger Settlement. This work has been of the nature of prospection and minor excavation of partricular younger structures. Radiocarbon dating is available for several of these and demonstrates their duration of occupation.

The full extent and complexity of these younger areas of Ferrandell-Oleza Settlement Complex and the structures they contain are not known, as the architectural elements stretch out over many hundred of meters in length and with an as yet undetermined width. Furthermore, recent surveys in the escarpment areas to the west and above, both, the Old and Younger Settlements have shown that these areas were also part of the Settlement Complex. Artefact and other materials have been excavated belonging to the full range occupational duration of the Settlement Complex.

With these facts in mind we will examine only those of known chronometric age in the Younger Settlement and the contextual data representative of the Younger Settlement are as follows:

Talayot *1

<<<<<<<<>>>>>>>

A series of radiocarbon dates are available for Talayot *1 (**T1**) which lies some 100 meters to the southeast of the Ferrandell-Oleza Old Settlement. The Talayot, approximately 8 meters in diameter, is on the outside in poor state of preservation, but its entrance and inner walls and one of its centre pillar stones are still in place and in a good state of preservation. This inner areas were filled to a depth of 3 meters and an excellent vertical stratigraphy existed for this Talayot in which several stages of activity were discernable (**Figures 8 - 9**). It is also contains the oldest dates so far available for the construction of the SFO-YS Talayots and may well be diagnostic of early Talayotic contexts throughout the site.

(**1**) Charcoal taken from crevasses in the bedrock and from beneath the foundation stones and base of the central pillar stone give us what is believed to be a date close to the structures construction. This date is Stratum 5: 2910 b.p. (**960 b.c.**) ± 100 years (**SFO-YS 30a**). A hearth zone for Stratum 4: 2540 b.p. (**590 b.c.**) ± 60 years (**SFO-YS 19c**) dates the occupation. There is also an abandonment date or a destruction date in Stratum 3: 2500 b.p. (**550 b.c.**) ± 40 years (**SFO-YS 18c**).

Talayot *2

<<<<<<<>>>>>>>

This Talayot (**T2**) is similar in construction and preservation to Talayot*1, about 8 meters in diameter with a height in the best

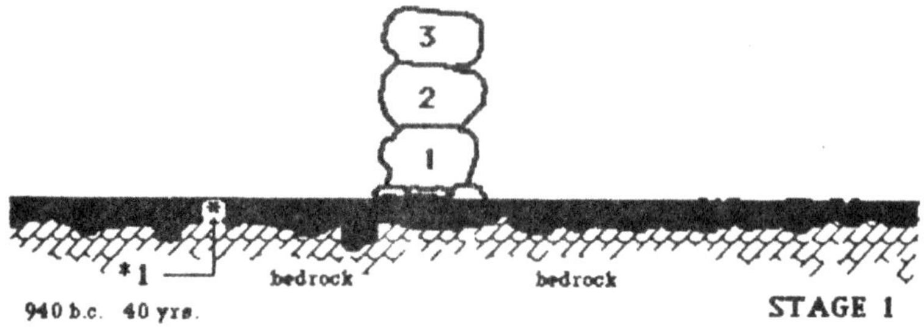

Initial Construction Phase. The uneven bedrock and general area are leveled off with material of the time and the three central pillar stones are erected. At this time, the charcoal and fires *1 were also deposited, becomi trapped in the fissures of the bedrock

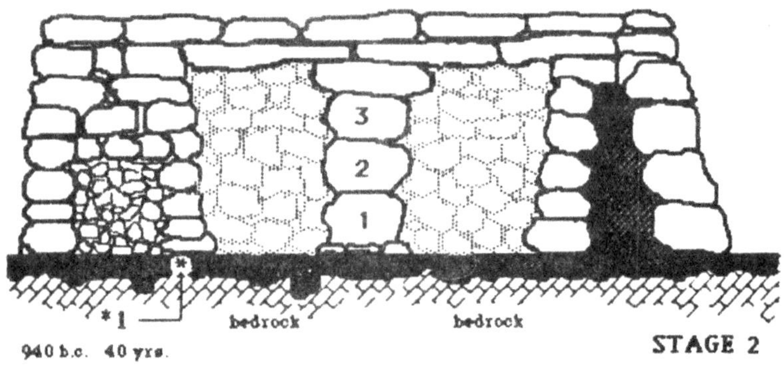

Full Construction Phase. The Talayot is completed and charcoal *1 and fire level are partly buried beneath the stones forming the entran and the central pillar stones. This is also the first occupation stage.

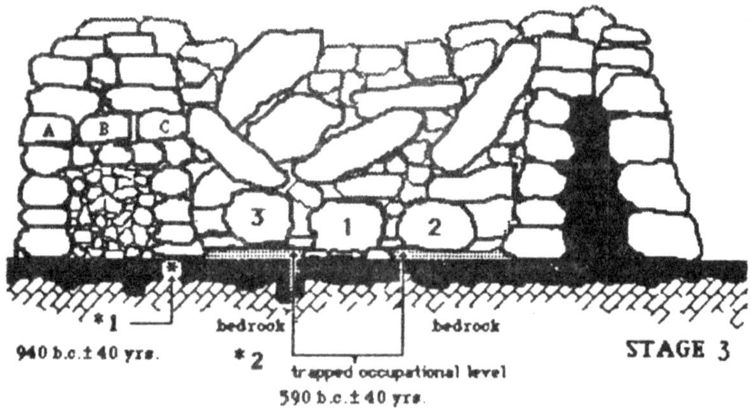

First Destruction Phase. The Central Pillar stones have collapsed and the roof of the Talayot has fallen in. The Entrance is still intact along with most of the walls. Datable occupation levels are trapped under the debris.

FIGURE

8

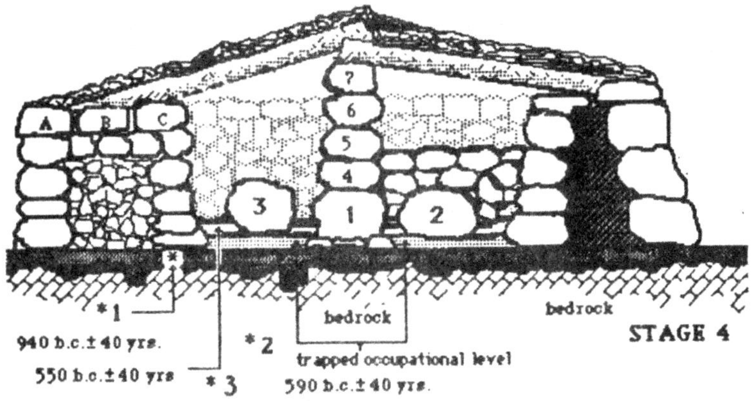

Renovation Phase. The debris of the First Destruction Phase is cleared away and a secondary pillar is raised incorporating the first element of the original Central Pillar. Numbers 2 pillar stone is incorporated in a dividing wall. At this stage the Talayot was probably roofed over with wood beams and covered with brush and earth. Occupation levels have accumulated, Stratum 3 has been dated.

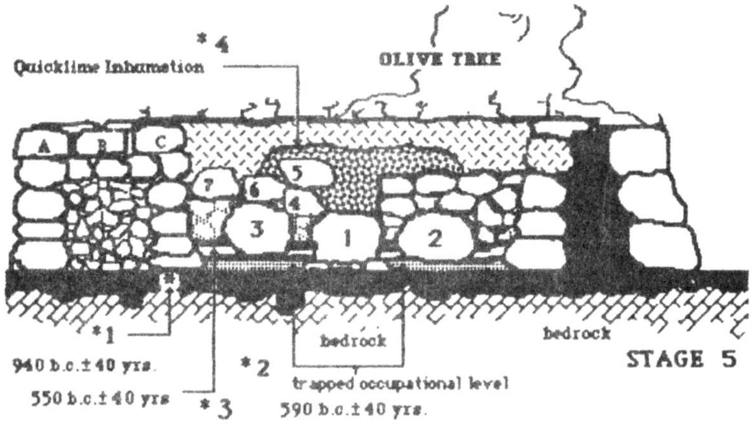

Second Destruction Phase. There appears to have been a razing and abandonment of the renovated Talayot (Stage 4). The smaller pillar stones 4-7have fallen and the centrl area of the Talayot has been used for a quicklime burial, as it contained human bones. It is at this stage that the excavations began.

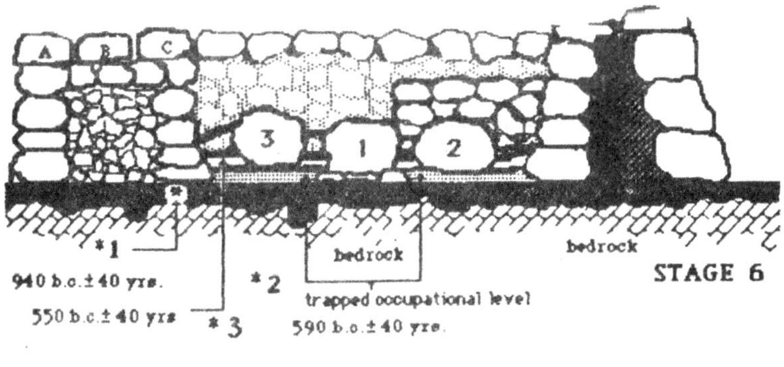

Final Excavation Phase. All the debris has been removed by excavation, exposing the the original levels, original Central Pillar stones. the dividing wall and smaller pillar stones. Pottery and charcoal from fissures in bedrock and beneath Central Pillar stone, *1, indicate a construction date of circa 1000 b.c.

FIGURE

9

preserved exterior section of over 4 meters. It has two of its centre pilar stones still articulated with a third and fourth fallen which formed the original four stone, high central pillar; although the entrance unlike T1 has been destroyed. Only half of the interior of this Talayot is excavated; however there are pertinent data available.

(2) One occupation date exists for Talayot *2 (**T2**) of the Younger Settlement in Stratum 3: 2460 b.p. (**510 b.c.**) ± 80 years (**SFO-YS 16b**). **An abandonment date is also available in Stratum 2:** 2400 b.p. (**450 b.c.**) ± 60 years (**SFO-YS 13b**).

Two work shop areas adjacent to Talayot 2 have also been excavated and can be dated relatively to the classical and indigenous artefacts found. These are the Tannery (H1) and the House of the Horses (**HH**), both with levels of circa 300 b.c. These are under separate study by the author. Radiocarbon samples have been collected for general correlation.

Talayot *3

<<<<<<<<<>>>>>>>>

This Talayot (T3) was found to contain no central pillar. It appears to have been uncompleted as several exterior stones are lined up on the exterior ready for construction. No radiocarbon dating is presently available from this structure. Artefacts from this Talayot are late in chronology, circa 200 b.c., and a separate study is underway at regarding the building's history.

Talayot *4

<<<<<<<<>>>>>>>

This Talayot (**T4**) has been recently, partially excavated and under study at the present time by a Reading University team directed by Dr. Robert W. Chapman and the author (**publication pending**). It has two central pilar stones still in place and is the the best of the four Talayots excavated todate; although its entrance like all the rest except T2 has been destroyed. A series of radiocarbon samples have been collected from pertinent contexts on the interior and exterior of the structure, but at present no readings are available.

Many excavated areas in both the Ferrandell-Oleza Old Settlement and Younger Settlement have produced radiocarbon samples that are pending. A series of 20 samples are at present in the pipe line and should available in the near future; even in addendum to this present publication.

Many of the artefacts themselves, especially from the Younger Settlement contexts, are such as to make dating of associated levels accurate within known '**classical**' chronological contexts, e.g. the presence of imported pottery and other items of known typological age. All the same, charcoal and other organics from such levels have been systematically gathered for future cross correlation and details concerning the first classical contacts. The diagnostic indigenous pottery found along side the imported items in these late contexts of the Post Talayotic Period compares very favourably with 14C dated contexts for the same type of artefacts found in the rock shelters and caves Post talayotic Iron Age stratigraphies.

As a final note, the reader should bear in mind the fact that the first classical contact and colonisation of the Balearics is a matter of historical record, beginning with the Carthaginian colonisation of the island of Ibiza in 654 B.C. and ending with the Roman colonisation in 123 B.C. Indeed, this kind of exposure to and involvement in classical military ventures and trade encompasses the whole of the Post Talayotic Iron Ages and undoubtedly includes regular contact with the areas of the Spanish Mainland and Southern France.

THE SECONDARY RESEARCH SITES

The Rock Shelter of Muertos Gallard and the Cave of Son Marroig (2, A, AMG and 2, B, SMRG) (6°,20'E, 39°,40'N)

Introduction

These two secondary research sites are considered together for two reasons: (**a**) the sites are geographically adjacent to one another, being separated by about 25 meters, and (**b**) they form a single sequence in regard to their radiocarbon chronology and artefact evidence. While they share proximity and age range, they differ in that one is a rock shelter and the other a natural cave. Both however, were also cemeteries during at least three periods of local prehistory and demonstrate four distinct burial customs. This is so for the Muertos Gallard rock shelter which is in essence a third or quarter sized, scaled down version of the Matge rock shelter.

In order of age, these burial customs at the two sites are the following:

Muertos Gallard:
(1) primary burial,

Marroig:
(2) secondary burial
Muertos Gallard:
(3) cremation
(4) inhumation in quicklime.

The utilisation of the area of the two sites for both living and burial is similar to that which took place in the Rock Shelter of Son Matge; the difference being that at Matge we have no evidence of Pretalayotic burial and only type three and four of the Muertos Gallard burial sequence is present. It is these aspects along with others at the Muertos Gallard/Marroig site that makes them of special importance as secondary sites. Another common aspect between the sites is the information and data supporting the (**NECP**) Early Neolithic Ceramic Phase of the Pretalayotic Period, in that both Muertos Gallard and Matge have produced the oldest recorded contexts for pottery in the Balearics.

In total there are 4 radiocarbon datings available for the following contexts:

(1) A radiocarbon date is available for an occupation level on the parapet area of the Muertos Gallard shelter overhang associated with indigenous pottery of Neolithic typology. The date is: Stratum 3, Meter Marker 11, Area of 'N' Rock: 4760 b.p. (**2810 b.c.**) ± 50years (**AMG 51a**). This date correlates very well with the Matge context with similar pottery typology 4650 b.p. (**2700 b.c.**) ±120 years (**ABSM 51**).

(2) An EBP Beaker burial associated with beaker incised wares comes from Meter Marker 12, Area of 'M' Rock in the form of a single grave (**Grave 1**)(**Valdren 1967 and Valdren 1982, plate 43:1 and Chapter V,Vol. II**) where a date of 3790 b.p. (**1840 b.c.**) ± 80 years (**AMG 45**) is in evidence. Here too, it is interesting to note the close correlation of this date for the Beaker contexts and those of both Matge and Ferrandell-Oleza Old Settlement.

(3) Evidence of a late phase of the Pretalayotic Period is present in the secondary burials associated with pottery of the Late Beaker Phase (LBP) of the Cave of Son Marroig. Here radiocarbon dating of human bone has biven us a date of 3470 b.p. (**1520 b.c.**) ± 80 years (**SMRG 40**). Once again the correlation between contexts of Matge and Ferrandell-Oleza supports the contemporaneity of all four stations.

(4) Dating of the lime conglomerate in the Post Talayotic quicklime inhumations, 2230 b.p. (**280 b.c.**) ± 100 years (**AMG 8**), also correlate perfectly with the Matge series. Todate however no available dates exist for the Talayotic cremation burial at Muertos Gallard, but there is no reason to assume that these will differ greatly from the Matge contexts.

The stratigraphical details as they existed in these sites were varied and details and discriptions of each of these can be attained elsewhere (**Valdren 1967 and 1982**).

THE TERTIARY RESEARCH SITES

Introduction

The tertiary research sites available for examination and discussion are numerous, as can be seen from the inventory list of Balearic stations with radiocarbon documentation. However only those with especially pertinent or with a good series of radiocarbon dates are discussed at present. To detail all the individual sites with only one or two dates would be a much lengthlier task than can be undertaken here.

Until now the settlement and other research sites examined have been from the geographic zones of the high Northern Sierras of Mallorca, whereas the four sites discussed below as tertiary sites are settlements located in the inland regions of the plains or flat-lands of Mallorca, Menorca and the small island of Formentera. Furthermore, they are representative of what exists in these areas, as well as being a geographic and site-type cross section. These tertiary sites are the following.

Mallorca:
(1) Talayotic Settlement of Son Fornes
(2) Talayotic Settlement of Ca Na Cotxera
Menorca:
(3) Settlement Complex of Torralba den Salort
Formentera:
(4) Megalithic Tomb of Ca Na Costa, Can Sargent

The geographical, stratigraphical and artefact contexts and other details involving all these sites can be and are best attained from excavational and interim reports. It will suffice here to briefly examine the radiocarbon documentation as a series form rather than the individual analysis themselves , and view what correlations if any these have with the primary or secondary research sites already examined.

Talayotic Settlement of Son Fornes

Perhaps the most important series for a Talayotic settlement todate comes from Son Fornes (**Gasull, Lull and Sanahuja 1984**). The site is located almost in the geographical centre as well as agricultural centre of the island of Mallorca, and on this basis is a model site to study settlement economy and social structure in the lowlands of the island; aspects reviewed in quite some detail by the excavators.

(1) This series along with artefacts and other evidence in the form of extensive statistical studies fully documents a portion of the upper part of the Talayotic Period from 2700 b.p. (**750 b.c.**) ± 170 years (**FORN 24a**) to 2450 b.p. (**500 b.c.**) ± 190 years (**FORN 16a**), although the statistical margins of error for the radiocarbon dates are high. Furthermore they suggest dates for a stage of construction and occupation of the site which are correlatable with those of sites like Ferrandell-Oleza and Torralba den Salort, where similar agricultural activities took place in prehistoric times even as they do at present. These are important aspects which I believe will become even more valuable and revealing as excavations of these sites progresses on them.

From the sites size and the complex nature and extensive building pattern that went on in the Fornes settlement, it is almost certain that earlier Pretalayotic contexts probably exist in the near vicinity.

Talayotic Settlement of Ca Na Cotxera

This site is best considered from the standpoint of its field excavation publication (**Cantarelles Camps 1972**). Its inclusion here is one of review and sequential order as an open-air settlement with an important radiocarbon date. Although, the site has only one radiocarbon date, it is included here because of the antiquity of the date and the materials associated with it.

(1) Inferior Level: 3750 b.p. (**1800 b.c.**) ± 120 years (**CI 47**). The site is also the first open-air site to produce datable materials associated with Beaker pottery; although a precise relationship with any Pretalayotic architectural is not present.

Talayotic Settlement of Torralba den Salort

In the case of the Settlement Complex of Torralba den Salort on Menorca, the radiocarbon series representing this open-air site demonstrates wider chronological perimeters than those of either Son Fornes or Ca Na Cotxera. The architectural and archaeological contexts studied there were a Talayotic Taula Sanctuary and its adjacent areas. Here dates from the interior of the Taula Santuary suggesting construction dates are the oldest found there.

(1) Adjacent to the Taula Sanctuary areas associated with a Naviform Talayot and a Pretalayotic house have produced a good range of dates. A date from carbonised

grain found on the interior of a Pretalayotic vessel from an abandoned house structure under the Naviformed Talayot gave a date of 3030 b.p. (**1080 b.c.**) ± 70 years (**TT 34b**) along with two charcoal dates from the same Stratum as the grain sample gave dates of 3020 b.p. (**1070 b.c.**) ± 60 years (**TT 34a**) and 2970 b.p. (**1020 b.c.**) ± 70 years (**TT 33a**).
(2) One dating is available for the exterior foundations of the Taula Sanctuary, Inferior Stratum: 2840 b.p. (**890 b.c.**) ± 30 years (**TT 29a**), another dating a Talayotic structure, probably a storehouse, gave a date of 2710 b.p. (**760 b.c.**) ± 50 years (**TT 25a**).

(3) Several dates are also availble for the inside of the Taula Sanctuary itself: Crevice in Bedrock Stratum: 2830 b.p. (**880 b.c.**) ± 40 years (**TT 29**), occupation dates of 2180 b.p. (**230 b.c.**) ± 100 years (**TT 5**) and abandonment of the Taula Sanctuary 2090 b.p. (**140 b.c.**) ± 50 years (**TT 3a**) make up the rest of the series.

The discovery of a few sherds of incised and puntated decoration gives a clue or indication to a still earlier chronology for the site; although no absolute dating of the contexts of these pieces as yet has been carried out.

The Megalithic Tomb of Ca Na Costa

This later site unique in its architectural typology (**Fernandez, Plantalamor Massanet and Topp 1974**) and the geographic area from which it emerges and has an interesting radiocarbon series, although unfortunately late for the most part.

(1) One date from human bone associated with the megalithic tomb, probably represents the last burials in the tomb and not its first or early utilisation. Some incised Beaker incised ware has been found with other artefacts of Beaker assemblage. The earliest date is: 3270 b.p. (**1310 b.c.**) ± 80 years and demonstrates once more a late context for what should be an earlier one. Other dates of 2670 b.p. (**720 b.c.**) ± 60 years (**CNC 23b**) and 2500 b.p. (**550 b.c.**) ± 100 years (**CNC 18b**) speaks for later activity in the area.

The late dates related to and associated with the Pretalayotic contexts in these tertiary sites are both interesting and revealing. This is so especially when they are correlated with one another. As we have seen, there are similar late dates of abandonment for the Ferrandell-Oleza Old Settlement and late construction dates of the Talayot in the site's Younger Settlement. This suggests the need for some change in our concept of the duration of the Pretalayotic Period, with some adjustments necessary for the upper interface of the Pretalayotic Period and that of the Talayotic Period. On the basis of this, adjustment has been made and the interface between the Late Pretalayotic Period and the Early Talayotic Period is considered circa 1300 b.c. Other minor adjustments down the line have also been made where there has been evidence.

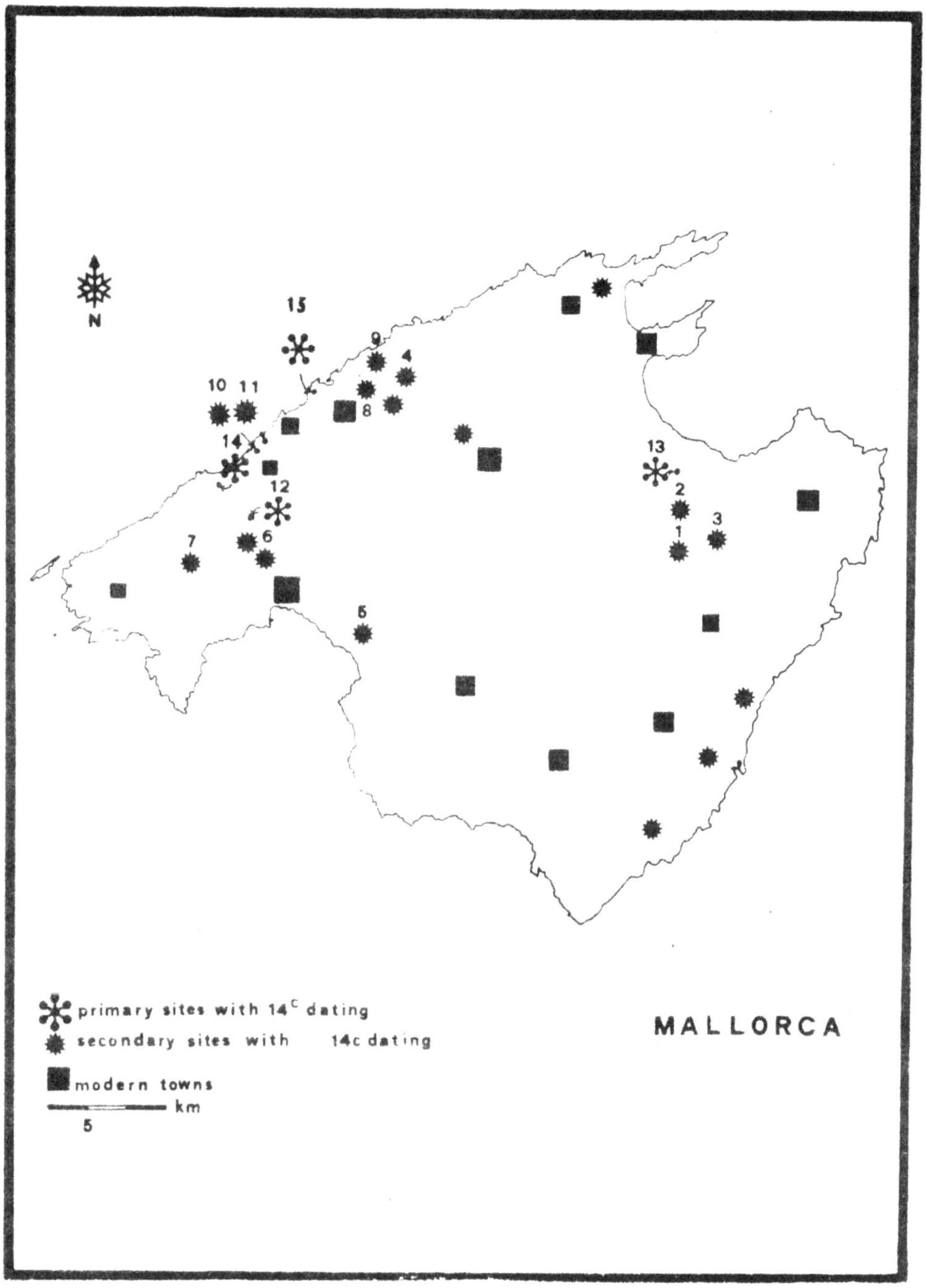

Map of Mallorca Research Sites with Radiocarbon Dates

Primary Sites

Muleta 15, Maige 12, Ferrandell-Oleza 14

FIGURE

10

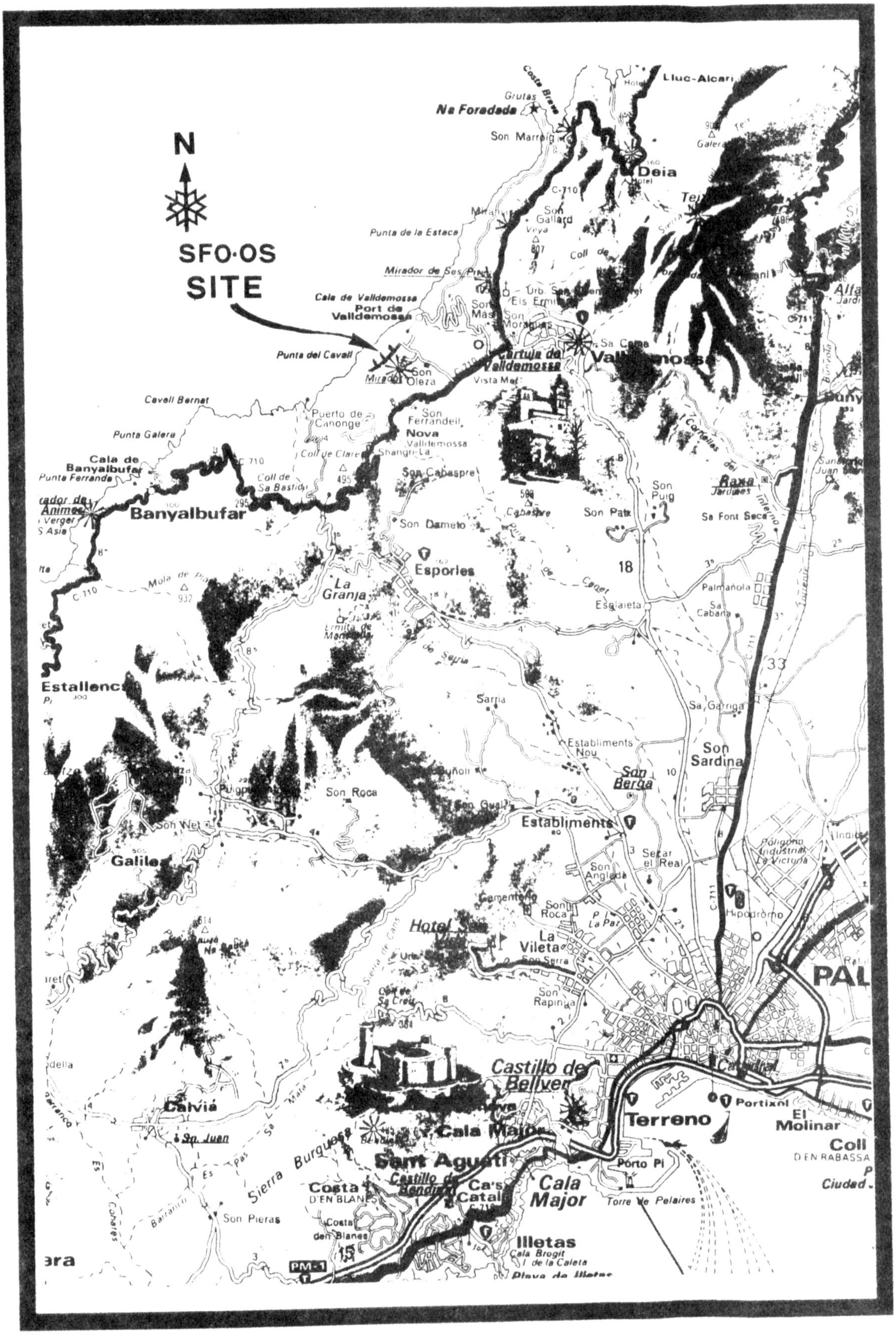

N
SFO·OS
SITE
Na Foradada
Deia
Valldemossa
Port de Valldemossa
Banyalbufar
Esporles
La Granja
Estallencs
Galilea
Establiments
Son Sardina
La Vileta
Castillo de Bellver
Terreno
Cala Major
Sant Agustí
Illetas
Calvià
Porto Pi
PAL
El Molinar

FIGURE

11

III

THE PENTAPARTITE DIVISION OF BALEARIC PREHISTORY

Introduction

In the general introduction at the beginning of this edition, a cursory outline of the author's chronological scheme has been given as a preliminary to this more detailed explanation of the different periods and their phases. Also, equally brief descriptions of the chronological frameworks of other investigators have been outlined below as a background to the chronology and general history of the investigations.

One need only consult the text books dedicated to the study of prehistory to be confronted by the fact that these new chronologies as well as the older ones are little known outside the Balearic Islands. Part of this shortcoming lies in the fact that these chronological schemes in general have been strictly local ones, concerned mainly with internal chronology, with the results that Balearic prehistory, apart from a little more than a decade ago, has been very much of a backwater; one where little attention has been given to its perspective in Western Mediterranean contexts, or in Western European contexts for that matter. It has also been vigorously emphasised that the use of radiocarbon age determination has now made it possible to correct this past scholarly neglect, and to present a chronology that is farther reaching and more precise than any proposed so far.

As a beginning, the old tripartite division of Balearic prehistory made up of Pretalayotic, Talayotic and Post Talayotic Periods, designed and used by the many investigators since the time of Cartailhac, Colominas, Bosch-Gimpera and others, is still a sound system in the author's opinion; and as can be seen, he still employs these nomenclatures as a nucleus to his system here. The main differences being that the author's pentapartite division has (**a**) the addition of two new periods, with Presettlement Period (**PSP**) and the Early Settlement Period (**ESP**) with (**b**) subdivisions added to the old tripartite cultural nomenclature of the Pretalayotic (PRT), Talayotic (**T**) and Post Talayotic (**PT**) Periods into Early, Middle and Late phases with a slight variation of these with the

Pretalayotic phases and, finally, (c) the use of the terms Copper, Bronze and Iron Ages related to the Pretalayotic, Talayotic and Post Talayotic Periods.

One of the guidelines of this research is the author's opinion that by concentrating on a few sites and dealing with these in detail, he can present a more detailed and coherent picture than could have been obtained by more superficial study of a greater number of sites. Regardless of the already ample length of this publication, much more could have been written. However, the existing text is in a sense an operation of taking stock and establishing general hypotheses, using so far as possible only information of guaranteed quality and not in trying to establish rigid correlative associations and geographic boundaries for Balearic materials and artefacts; although possible parallels and comparisons are attempted and suggested with those farther afield, such as with the author's '**arc of immediate influence**' (**Part IV**) as related to the Balearic Islands throughout their prehistory. The new evidence, in the mind of the author, then necessitates a period of digestion by others, and the approach presented here should permit that important process to take place; all aims still valid and applicable at the updating of this work here in 1986. In this way, the author can draw a line to continuing in-coming data and information from his current excavational activities to the end of 1985, for the sake of presenting additional and the latest results so far, even though investigations are still continuing and may by the republishing of this work have already passed beyond that line.

History of Archaeological Research in the Balearic Islands

The first historical reference to the prehistoric people of the Balearic Islands was made by Diodorus in a text related to the year 406 B.C. concerning the recruiting of Balearic mercenaries by the Carthaginians for their campaign on Sicily. Other references have been made by early historians and travellers such as Strabo and Pliny.

The first mention of the monuments in general histories occurs in 1653 by Binimelis who spoke of their magnificence, attributing them to the works of giants and demons. In 1752, J. Armstrong published his **History of the Island of Menorca**. The first prehistoric Spanish study appeared in 1818 by Ramis y Ramis, associating the monuments with the Druids; 1844 saw the formation of the first Archaeological Committee for the Balearics dedicated to the preservation and study of sites. The first classic treatise on megalithic architecture was written by J. Fergusson in 1872. This was followed in 1880 by the formation of the Luliana

Archaeological Society in Palma. The French scholar E. Cartailhac drew together earlier studies on a scientific basis in 1892 in the publishing of his **Primitive Monuments of the Balearic Islands**.

The first decade of this century began with an acceleration of interests in the Balearics. L.C. Watelin published the first relative chronology based on typology; A. Mayr put together his catalogue of known remains.

In 1920, decisive steps were made to systemise investigations by creating the Archaeological Excavations Service under the direction of Bosch Gimpera. Excavations under the direction of J. Colominas and J. Malverti went on for some years, during which time Bosch Gimpera was at work defining chronology and relationships in the Mediterranean.

1925 and 1934 were years in which several foreign scholars such as Chamberlin, Hemp and Kessler, became interested in the prehistoric problems of the Balearics. Margaret Murray's Cambridge Mission to Menorca, excavating at Trepuco and Torreta were exemplary; published in 1932 and 1939.

Most work was suspended until 1950 because of the Spanish Civil War and World War II. The direct result of a visit in that year by the IV Curso of Ampurias gave renewed impetus and interest to current insular studies. Since then National Excavation Services have been active, along with several foreign missions (two American, a German and an Italian), the William Bryant Foundation under the direction of D. Woods, the Deya Archaeological and Research Centre, directed by the author, the German Mission under the supervision of H. Frey and the Italian Mission directed by G. Lilliu.

Local archaeologists as well as directors of foreign missions have been brought together during the last two decades under the direction and collaboration with the Museum of Mallorca, supervised by G. Rossello Bordoy. The list of independent local workers during these years include excavations done by L. Amoros, B. Ensenat Estrany, B. Font Obrador and J. Mascaro Pasarius. Many eminent Spanish scholars can be added to the list of investigators such as A. Arribas, M.L. Serra Belabre, G. Bellido, J. Maluquer de Motes, P. de Palol, L. Pericot Garcia, J. Crusafont Pairo, L. Pantalamor Massanet, M. Fernandez-Miranda, M. Tarradell and C. Veny.

The author's involvement with Balearic prehistoric problems began in 1960 with the discovery of the cave of Son Marroig and the rock shelter of Muertos Gallard. At that time the author was encouraged to follow his investigation by J. Malverti and L. Pericot Garcia. These sites were followed by the discovery of the cave of Muleta in 1961 and finally the rock shelter of Son Matge in 1968. Since then his other excavational activities include the current co-directorship with M. Fernandez-Miranda of the Talayotic Taula of Torralba den Salort on Menorca and directorship of the Talayotic and Pretalayotic Settlement Complex of Ferrandell-Oleza in Mallorca; and in 1984 the co-directorship with C. Morell Orlandis

During the last two decades a number of different chronological schemes have been presented by investigators, including the latest one proposed by the author. The most important of these have been proposed since 1960 by investigators such as G. Lilliu (**1962**), C. Veny (**1968**), G. Rossello Bordoy (**1973**) and M. Fernandez-Miranda (**1979**). Apart from certain points of nomenclature and chronological boundaries, these recent frameworks do not differ radically from that set down by E. Cartailhac and J. Colominas Roca, who introduced the name '**Talayot**' into the scientific vocabulary for the use in describing the local Balearic culture. For the most part, the tripartite division of Balearic cultural prehistory, consisting of Pretalayotic, Talayotic and Post Talayotic Periods has been generally accepted, although the last decade has seen investigators attempting to refine this tripartite system.

The main stimulus for change in thinking concerning Balearic prehistoric chronology has been the introduction of the use of radiocarbon age determination into the Balearic Islands. Since the author introduced the use of this method into the interpretation of the stratigraphical horizons in his sites (**Waldren, 1965, 1966, 1967, 1968, 1970, 1972 and 1980**), the method has revised many views regarding Balearic prehistoric problems (**e.g. Pericot Garcia 1973, Rossello Bordoy and Waldren 1973; M. Fernandez-Miranda and Waldren 1976 and Fernandez-Miranda 1979**). Among other recognition, an important critique regarding the success of radiocarbon analysis as applied to Balearic problems along side Mainland Iberian contexts also has been recently reviewed (**Fernandez-Miranda 1978 and Almagro-Gorbea 1978**).

A Summary of the Current Chronological Schemes

As this introduction deals with initiating the reader in past as well as present research in the Balearics, a brief summary of the various chronological schemes used to divide the local prehistory should illustrate the immediate differences between each of the major frameworks presently used in interpreting local prehistoric events. Accordingly, these may be compared in the text with regards to their divisional nomenclature with the author's chronological scheme, outlined earlier in this introduction . It will be seen that while the introductory use of radiocarbon dating has been responsible for the formation of the more recent of these new chronological schemes- which from the point of view of actual chronology differ from one another very little- they do little to compare, suggest parallels or, for that matter attempt to correlate the Balearic data and material evidence with the wider perspective of Iberian prehistory, or with the larger Western European context.

The main criticism the present author has with these local new chronologies, as well as those in the past, has been their tendency to regard the Balearic problem as a strictly insular one. While this is a safe position to take, and while there are many of the local developments that can be considered solely insular, there are many material signs of influence and exterior reflections of events that are the products of outside trade and contact from within an '**arc of immediate influence**', the location of which is mainly western, extending from southwestern to eastern Iberia into the Pyrenees and Provence; a geographic arc of influence, the direction of which existed to a large extent during most, if not all, Balearic prehistory (**see Part IV**).

The chronological schemes outlined below represent the most serious of those proposed in the last twenty years. The first is the result of excavation at the fortified settlement of Ses Paisses on Mallorca. It was excavated by G. Lilliu from 1959 to 1962 and his hypothesis for a chronology is based on a localised stratigraphical interpretation and one radiocarbon date, obtained from a series of living quarters within the village compound; the chronological plan being based on the stratigraphy of two particular buildings (**House 12 and House 3**) in which he recognised three distinct layers, as follows:

Talayotic I	**House 12**	14c Date: **950 b.c. ± 100 yrs.** (on bedrock)
	House 3	**earlier than 800 b.c.** (on bedrock)
Talayotic II		**circa 700 b.c. to circa 400 b.c.**
Talayotic III		**circa 400 b.c. to circa 200 b.c.**

The second of the new chronologies is proposed by G. Rossello Bordoy (**1973**) in which this investigator bases a good part of his chronological framework on recent radiocarbon dating but, also proportionately, on his excavations and experience in the field. This framework and the following one, except for chronological details, are quite different from the author's framework; with large differences in nomenclature and subdivisions. Bordoy's chronological scheme consists of five divisions, which are as follows:

Pretalayotic	circa **5000 b**.c. **to** circa **1400 b**.c.
Talayotic I	circa **1400 b**.c. **to** circa **1000 b**.c.
Talayotic II	circa **1000 b**.c. **to** circa **800 b**.c.
Talayotic III	circa **800 b**.c. **to** circa **500 b**.c.
Talayotic IV	circa **500 b**.c. **to the Roman Colonisation**

The third and latest of these new chronological schemes is proposed by M. Fernandez-Miranda (**1979**) and consists of the following five periods.

I.	**Preceramic Man**	circa **5000 b**.c. **to** circa **2000 b**.c.
II.	**Incised Pottery Horizons**	circa **2000 b**.c. **to** circa **1700 b**.c.
III.	**Bronze Pretalayotic**	circa **1700 b**.c **to** circa **1500 b**.c.
IV.	**Talayotic I**	circa **1200 b**.c. **to** circa **800 b**.c.
V.	**Talayotic II**	circa **800 b**.c. **to** circa **100 b**.c.

In the author's opinion, these chronological frameworks are cautious ones, which deal mainly with local prehistoric events, even though more than ample chronometric data and materials are available for more detailed schemes; the proposal of such a more detailed scheme being one of the main objectives of this present hypothesis. Therefore, it is the author's contention that with the present 135 absolute dates for Balearic prehistory, and with the excellent existing stratigraphies and their materials from such sites as Muleta and Matge and the many other Balearic stations, a more ambitious chronological framework should be possible. One which ventures to examine and suggest possible external parallels, correlations and relationships with mainland sources of influence and contact and not, mainly, with comparative internal sites about which a great deal has already been said. Besides, the homogeneous nature of the local sites and materials, though quite remarkable in themselves, are equally rather monotonous as they are presented up to date, by merit of their lack of a wider perspective and context. Therefore, a need to venture farther afield than just the Balearic Islands themselves is quite imperative.

The following text will be divided into sections devoted to the description of the chronological periods and the evidence from the three primary research sites; beginning with the Cave of Muleta and followed by that of the Rock Shelter of Son Matge and Settlement Complex of Ferrandell-Oleza. The sequential order of these sites is a natural one, not only of their order of discovery, but also a chronological one; Muleta being not only the first site to have been discovered but the station with the oldest stratigraphical material evidence.

Therefore before the description of the chronological framework formed by the physical and other evidence from the Matge rock shelter or the Ferrandell-Oleza settlement and their stratigraphy and chronology, there will be a section on the period of presettlement dedicated to the evidence available from the Muleta deposit and a summary of the various research studies presented in greater detail in the Appendices of the author's thesis (**Valdren 1982, Volume III**). The inclusion of this text at this particular point in the text is done in order to present details on the important ecological evidence of the Presettlement Period, which is only found in the Muleta deposit and in one of the other research sites. In this way, as this ecological information relates only to the Presettlement Period and Muleta, no further description of this material and evidence need be made elsewhere in the general text. At the same time, it prepares the reader with environmental and ecological background in the description of the evidence which forms the structure of the Early Settlement Period and its presence in the Rock Shelter of Son Matge; the description, stratigraphy and chronology of which was the subject of Part II.

The **secondary** sites of Muertos Gallard, Marroig have been included and described briefly at the end of Part II. The descriptions and stratigraphies of these minor or secondary sites have been treated in a cursory manner, mainly, because their physical stratigraphies and chronological horizons are, as explained earlier, by far less complicated and narrower chronological range than those of the **primary** or major sites of Muleta, Matge and Ferrandell-Oleza; the secondary sites consisting of only one or two stratigraphical or chronological contexts and by far smaller amounts of artefact evidence.

The '**chronological scheme**' that follows provides the reader with the description of all the chronological periods, the phases and physical and other evidence. The description and frame work below not only presents a more detailed accounting than the summary found earlier in General Introduction, but also delineates the chronometric analyses results available for each period and their phases. As is found in Part V, a complete inventory along with tables of conversion for Bristlecone-cone Pine calendar years and conversion tables of Balearic radiocarbon dates is available, along with other chronometric studies, where the dates appear in individual tables and which the reader can refer back to at any time while reading the general text here.

I

THE PRESETTLEMENT PERIOD

<<<<<<<<<<<<<<<<<<<<>>>>>>>>>>>>>>>>>>>>

fossil record until circa 5000 b.c.

The beginning of this period opens with the earliest evidence, so far, in the fossil record of the artiodactyl, ***Myotragus***, of which three currently identified species exist. There are some very recent findings which suggest that there were more species than these three types, but these new species will have to be excluded from this edition pending future publication. However, a tentative family tree in the form of a list is provided below to give the reader some idea of the genealogy. The three currently accepted species are (a) ***M. Antiquus*** (b) ***M. batei*** and (c) the ***M. balearicus***.

In our study of the Presettlement Period most of the research is related to the species ***Myotragus balearicus***. A brief discussion of the M. antiquus and ***M. batei*** species has been included here in order to show the major evolutionary developments in the physical structure of ***Myotragus*** and, more specifically, that manner in which particular changes in its morphology possibly point to changing ecological conditions. While these aspects are yet not as fully documented in the fossil record as palaeontologists may wish, and as will undoubtedly become available in the future, the subject is approached in order to demonstrate the odd evolutionary path of one insular species by means of the fossil record. It will also illustrate how development, diagnostic of insularity, may proceed when there is a limited number of indigenous species, and how extinction can occur as a combined result of over-specialisation and contact with man. This latter philosophical lesson often bet hidden under the weight of more statistical and scientific approaches.

The Balearic Islands are typical of other islands in the Mediterranean as well as elsewhere in that, as a general rule, islands are well known for their restricted indigenous fauna (**Sondaar 1973**). The species that manage to exist usually run to highly specialised types of a limited variety. For example, in other areas of the Mediterranean, dwarfing and giantism at times as with the giant dormouse ***Hypnomys morpheus*** (**Bate 1918 and Mills 1974**) among the extinct insular fauna is common (**i.e. dwarf elephants and rhinos are known from Crete and Cypress: Boekschoten and Sondaar 1973**). Deer are also common in the fossil record of most of the Mediterranean islands; however, the Balearics lack this species in its fossil record, and the only other island known to have an antelope is Sardinia (**Dehaut 1920**).

Dorothea Bate of the British Museum discovered the first remains of ***Myotragus balearicus*** (**Bate 1909**). Although she did not find evidence of man associated with ***Myotragus***, she did scientifically describe the peculiarities of this indigenous animal along with the **Hypnomys** and **Nesiotites**; and hence, she can be said to have been the first prehistorian to exemplify locally the fact that islands can be rare and valuable storehouses of palaeo-ecological data, information and materials. She was to retain keen interest, not only in ***Myotragus*** but also a wide interest in other insular Mediterranean species until her death in 1946.

Here, apart from the work of a small number of local geologists and palaeontologists, interests in the ***Myotragus balearicus*** remained until the discovery of the cave of Muleta in 1962.

Until the discovery of Muleta, the overwhelming majority of remains of ***Myotragus*** were encountered in well consolidated cave '**breccia**' sediments of Pleistocene origin, generally accredited to a period some 40,000 to 20,000 years ago. For the most part, these earlier remains consisted of highly fragmentary bones which were difficult to remove from these deposits.

The discovery of Muleta cave with its Pleistocene and Holocene deposits of extraordinary rich and well preserved macro-microfauna, especially the ***Myotragus*** remains with their complete '**mint condition**' ossious bone components, has been not only instrumental in re-awakening Balearic prehistoric ecological interest and quaternary research in the area in general, it has also produced human remains associated with ***Myotragus balearicus***, thereby demonstrating a possible symbiotic relationship between the two and, at the same time, adding a completely new dimension to prehistoric studies in general.

By closely studying these ***Myotragus*** remains with their osteological and morphological peculiarities, reflected in such aspects as tooth development, competitive behaviour and sexual dimorphism in the animal, it should be possible to establish some hypotheses. Hypotheses not only about ecological conditions of the time but also to what degree such evolutionary developments and adaptations to those conditions might have had concerning man's relationship with ***Myotragus***. For surely all these aspects would have had some, if not great influence on man's treatment of the animal during such pursuits as hunting, herding or other subsistence oriented activities.

Some of the peculiarities and dramatic adaptation of insular species have been pointed out in subsequent sections, where the reader has been alerted to such insular adaptation as dwarfism (**e.g. island rhinos and elephants**), and even giantism as in the case of the Balearic dormouse, ***Hypnomys***. However none of these developments, as dramatic as they are, compare to the odd and different evolutionary changes and adaptations undergone in the body structure of ***Myotragus balearicus*** in its insularity.

On the Origin of the Species

<<<<<<<<<<<<<<<<<<<<<<<<<<>>>>>>>>>>>>>>>>>>>>>>>>>>

While we have a great deal of physical and chronometric evidence of its late survival and even the final extinction of ***Myotragus balearicus*** and although recent amplification of its early fossil record has done much to give us a relative idea of its age, little is known of its origin. There are however several different opinions as to when the ***Myotragus*** first arrived on the islands. To simplify the discussion of these various hypotheses the following list has been made, and while they are brief accounts of these different hypotheses, they contain the bases of the various ideas as to the origin of the ***Myotragus***. The four hypotheses are as follows:

(1) The Late Miocene to Early Pliocene (Colom 1975)

G. Colom believes that Mallorca during the Late Miocene to Early Pliocene had a major contingency and variety of biotas, and argues that ***Myotragus balearicus*** could have easily passed from Mallorca to Menorca during the Pliocene, along with other species (which is rather disconcerting as he does not postulate what other animals). However, he does believe that the reason the animal did not migrate from Mallorca-Menorca to Ibiza and the Iberian Mainland is the island of Ibiza was already separated from the Mallorca-Menorca land mass.

(2) The Late Pontain, i.e., minimum of 5 to 6 Million Years Ago (Cuerda 1966)

J. Cuerda argues that the sea levels of the Late Pontain Epoch were some thousand meters lower and therefore what is now Mallorca and Menorca would have been incorporated in the Continental land mass, and free access to a migrating ***Myotragus*** ancestor could have found itself on the particular area which was to become Mallorca and Menorca. He does not explain the complete absense of Myotragus on Ibiza. For surely had Myotragus migrated from the direction of Mainland Iberia, we would have fossil evidence on the Island of Ibiza by now for that event.

(3) The Early Pleistocene, Late Villafranchian, circa 2.5 Million Years Ago (Crusafont Pairo 1966)

M. Crusafont Pairo's assessment is made on the basis of the ***Myotragus*** batei which until recently was the oldest species of ***Myotragus*** found in the fossil record. This highly intelligent estimate seemed reasonable on the evidence of 1966. However, here in 1986, we now know there are other species of ***Myotragus*** which on evolutionary evidence alone are certainly older.

(4) The Mindel Glaciation, estimated at 6,000,000 to 5,000,000 Years Ago (Alcover y Tomas 1976)

Alcover y Tomas argues this origin on the basis of continental fauna of this period such as the Gallogoral, stating that the ***Myotragus'*** ancestors arrived in the Balearics from Eurasia before the Pontain Epoch, and probably about the time of the Mindel Glaciation; which in the present author's mind covers a large area of postulation.

In seeking an assessment as to when and from where the first ***Myotragus*** arrived in the Balearic Islands, it is the author's hypothesis that this event must have taken place by the Late Pliocene or Late Pontain, 6 to 8 million years ago. This assessment is based on the author's consideration of the existing fossil evidence of at least three species of ***Myotragus*** involved to date. It also takes into account the ***Gallorgoral meneghinii*** of Western Europe and the ***Antelope meloni*** of Sardinia. It is an established landmark in the fossil record of Eurasia that the first antelopes began their radiation southward by the middle of the Miocene (**Romer 1961**) and had by that time developed into at least a half a dozen different species by the Late Miocene, already on the move to the south from Eurasia, it is not unreasonable to postulate on evolutionary grounds seen through the species ***Myotragus***, that the earliest type may well have arrived in the Balearic region by the Late Pontain or Late Pliocene, thus becoming isolated shortly afterwards. It is quite apparent that we are dealing with an insular genus in ***Myotragus***, which most assuredly remained insular during its evolution, resulting in the three or more species we know to date. That we will eventually find other species or sub-species is, in the mind of the author, most probable. It is also very evident that the case of the Antelope meloni (**Dehaut 1911**) is another result of insularity, though the argument for comparison with ***Myotragus antiquus*** grows stronger with each new discovery. If it should prove to be the case that the ***Antelope meloni*** is the closest relative to ***M. antiquus*** or still earlier species, then some very extraordinary geological implications should be forthcoming, in that relationships between the once island mass of Mallorca-Menorca with the island of Sardinia must have taken place.

In 1977, the author and P. Sondaar of the University of Utrecht, Holland were invited to collect additional bone-bearing rock from J. Pons' Cap Farrutx Mallorcan site. The bone bearing matrix consists of sediments of very old appearance in fissures of Miocene limestone, taken from an elevation which is one of the highest in the region. This material has since been sent to the University of Utrecht in Holland, and treated there to extract the bone specimens. Study of these specimens has not yet been completed, but preliminary examination suggests that a fuller and perhaps in some way modified description of ***M. antiquus*** should eventually be possible, along with substantial additions to the meager knowledge currently available.

However, the emphasis of research concerning the evolution of ***Myotragus*** in this hypothesis really falls on the ***balearicus*** species of the Late Pleistocene and Holocene deposits from Muleta and Matge. It is concerned with the animal's final development, its relationship with man and its ultimate extinction during the Pretalayotic Period. As some reference has been already made to the role of insularity in the morphological development of the animal, we can now turn to the description and modifications that took place in the animal and compare them to other animals.

Myotragus balearicus (**Bate 1909**)

<<<<<<<<<<<<<<<<<<<<>>>>>>>>>>>>>>>>>>>>

The ***Myotragus balearicus*** was described at the beginning of this century by Dorothea Bate (**1909**) and Charles Andrews (**1914**) as a rather peculiarly developed and highly modified ***Nemorhaedine Antelope***, endemic to the Balearic Islands of Mallorca and Menorca. The animal's endemic peculiarities and modifications are found in its skull, jaws and fore and hind limbs. There are many other characteristics which are described in detail in the appendices of the author's thesis (**Waldren 1982: Appendix 1D**). However, these are too numerous to include here in the text, and are set down in the appendices there for the interested reader. At the present, we are only considering the major osteological developments and the possible characteristics and conditions they suggest in the animal and its surroundings.

One of the most notable of its diagnostic characteristics is the extreme frontal position of its orbits. Such a development would have given the animal a limited angle of vision of about 110 degrees and stereoscopic vision and, at the same time, would have made the animal easy prey for a carnivore, had there been one (the ***Myotragus*** had no known predators, apart from man, as we will eventually see). This characteristic suggests that as it did not have predators and that this modification was an adaptation to its surroundings; and one with a particular function or even set of functions. Such stereoscopic vision would have been advantageous in the search of food between narrow crevices of rocks, or in judging distances in jumping from crag to crag in the mountains; both of which it would have had to do to survive even in the gentler conditions of the present-day mountainous regions of the islands. For that matter, its limited angle of vision would have made ***Myotragus balearicus*** easy prey for either man or other predators, as an animal with such a handicap, living under normal conditions with the presence of predators, would have been easy to approach from the rear without alarming it. Coupled with other osteological peculiarities, viewed shortly, these would have made the animal easy to hunt, and certainly would have shaped man's treatment of such an animal. An adaptation of this sort would also have affected its performance and behaviour during its mating ritual.

The second major modifications are found in its jaws and teeth. Most artiodactyls have three pairs of normal incisors in their mandibles. ***Myotragus balearicus*** in evolution lost two pairs of incisors and a set of premolars and the incisors have been replaced by a single robust evergrowing chisel-like incisor, similar to that of a rodent (hence its name, ***Myotragus***, mouse goat). These teeth are the main basis for distinguishing between the different species of ***Myotragus***. The number, shape and other characteristics of these teeth have changed during the evolution of the species, one to the other, and in so doing, suggest different functions and even environments themselves.

The evergrowing chisel-like incisor of ***M. balearicus*** is undoubtedly a specialised adaptation; one which certainly reflects the nature of its food, and might well suggest the types of vegetation and climatic conditions of the time. An incisor of this kind would be especially effective in either excavating for roots, scraping bark off tough vegetation or trees and even turning over rocks in search of lichens. The possibility of such activities are further supported by its steep angled jaw. In turn, a jaw of this sort would have to be supported by powerful neck muscles as well as powerful shoulders, all of which ***Myotragus*** had as demonstrated by the osteological evidence. Again, these characteristics are aspects that man would have taken into account in such situations as hunting, herding, etc. In short, these cranial and post-cranial characteristics suggest specific functions as well as conduct and environmental conditions, such as geographic and geological habitat and even dietary needs and the food sources themselves.

The third of the major modifications are found in the animal's fore and hind limbs, especially the metapodials (***metacarpals and metatarsals***). There is a general foreshortening in all of the bones of the limbs, though this is particularly noticeable with the '**cannon bones**' or metatarsals and metacarpals, which appear both short and stout when compared to those of deer, sheep and goat. Perhaps more than any bone, these show the great variability of bone sizes found in the full grown males and females, characteristic of the species (**Waldren 1982: Appendix 1D**).

The metatarsals of ***M. balearicus*** are the most interesting of the cannon bones in that the tarsal components, found as separate pieces in most other genera in order to give a lot of mobility in the hind limb, are in ***Myotragus balearicus*** fused together to form a single tarsal component. This in turn is frequently fused to the proximal end of the cannon bone, forming a tarso-metatarsal unit, diagnostic of the ***Myotragus balearicus***. Such a characteristic could only result in the '**slow locomotion**' (**Sondaar 1973**), and an inability to '**veer off course**' like other animals when pursued by predators. Even though we are more or less sure that ***Myotragus*** had no predators, and this could have been true also concerning the other ***Myotragus*** species, and it is not certain that this fusion of the tarso-metatarsal components took place in these others. Nevertheless, these modifications in the hind limbs in particular are strong indications that the ***balearicus*** species was a slow mover, and this is also consistent with the lack of a wide angle of vision; both of which would have been advantageous in a mountain dwelling creature, where judgment of distance and good balance are extremely necessary. It seems reasonable to say that there probably were many interrelated evolutionary factors involved in each of the animal's peculiar characteristics and that the '**slow locomotion**' feature of the animal's short and stout cannon bones is only one which would have stood it in good grace as to stability; another advantage in both climbing and jumping from mountain crag to mountain crag.

The metacarpals, which are smaller, shorter and stouter than the metatarsals, have carpal units which are independent as in other species and, so, would have assured good mobility in the fore limbs. As pointed out by C. Andrews (**1914**), this feature of short-stout metacarpals can also be found in the present-day Rocky Mountain goat of North America, and that in this species such features are a requisite to its survival and habitat. In ***Myotragus***, this feature could have been equally advantageous in its habitat where jumping would also have entailed catching itself on its fore limbs and absorbing the shock of landing prior to its hind limbs coming into play where stability became necessary.

There are a number of observations which can be made concerning the osteological evidence of ***M. balearicus*** from the Muleta deposit which are not skeletal modifications but which are physical conditions present in some of the remains. For example, while there appear to have been very little visible signs of disease '**per se**' in the animal, there are some very good examples of pathological conditions, such as bone trauma. There are a number of examples of healed broken bones, some times quite grotesque and which had there been predators would have never reached the age that they apparently did. However, all these are rare considering the number of animals we are dealing with in the Muleta deposit and the fact that sick or dying animals frequently seek the shelter of caves.

For the moment, the further subdivision of the Presettlement Period must depend on the relative chronology of each of the above three different ***Myotragus*** species. Accordingly, there follows in the next chapter a more detailed description of each of the three species and the times in which they lived, as viewed by the author and others through some of the dental and osteological and morphological adaptations undergone by the ***Myotragus*** and its evolution.

The evolutionary and morphological changes all come together in the species ***Myotragus balearicus***, in the Holocene epoch, ending circa 2200 b.c. It is rare when one can demonstrate such a long species -especially an insular one- and trace them up to the time of its extinction, as in the case of ***Myotragus*** in Muleta and Matge (**Waldren, forthcoming publication**); an essential requisite for any overview or treatment in general of Balearic ecology and culture. At the same time, the present data, information and materials, so to speak, sets the stage for a major study in the processes of insular evolution and prehistoric ecology and culture, as well as sets a course for future cave and rock shelter studies in general, along with perhaps increasing the possibility of still earlier discoveries of man in the Balearics by emphasising the type of materials and information available in insular areas. The importance of such geographically isolated areas as laboratories and warehouses of materials bearing vital data, information and materials valuable to many aspects of prehistoric studies is becoming apparent (**Evans 1973**).

For the moment, however, it must be remembered that the present picture gives us only a discontinuous series of glimpses of events, and environments, seen with a rather poor degree of focus. The deposits which should yield some of the future definitive evidence, the lower levels of some of the research sites, are only recently discovered and excavation on them only just begun. Nevertheless, the data and materials already available from them enables us to make a dynamic review of certain finds made much earlier in this century, and to begin to realise their implications. It seems well worthwhile, accordingly to present the current working hypotheses and tentative conclusions, but the reader must not look for the kind of detail and precision to which we can aspire during the various later stages of the prehistoric periods in the Balearic Islands.

At the present the chronology of the Presettlement Period in the Balearics begins with the fossil record some time in the Upper Pleistocene, about 8 to 6 million years ago, and extends into the Villafranchian epoch of the Lower Pleistocene, some 2.5 m.y.a. While the evidence and remains directly pertinent to our research do not go back so far, the true fossil (**petrified**) remains are offered here in descriptive form as a background of the fossil record and the relative age of these species-types studied here as part of the period. The following ages are assigned to ***Myotragus*** fossil types.

(1) M. antiquus	**5-6 m.y.a.**	**Upper Pliocene** (Pons 1977)
(2) M. batei	**2.5 m.y.a.**	**Villafranchian/Upper Pleistocene** (Angel and Crusafont 1966)
(3) M. balearicus	**1 m.y.a. to circa 2200 b.c.**	**Middle Pleistocene/Holocene** (Bate 1909)

The other mammalian fossil species indigenous to the Balearics are the rodent (**giant dormouse**), ***Hypnomys***, briefly mentioned above, and the insectivore (**giant shrew**), ***Nesiotites***. Together with the the ***Myotragus***, they form the '**trilogy**' of mammalian forms which demonstarate the restricted nature of islands and their capacity to sustain only small faunal populations. From this, one conclusion seems apparent from the outset, and this is that we can be reasonably sure that the ecological conditions of the islands were never such that they could support more than the limited number of animals present in the fossil record of any particular time.

In turn, it is now known (**see below**) that each of these species have fossil species-types related to a ***Myotragus*** species and correspond to a chronological division of the Presettlement Period. It is also apparent that each of these had their individual morphological characteristics (**Reumer 1980 and Mills 1974**) which can be accredited to a broad but given time.

The following ages assigned the ***Hypnomys*** and ***Nesiotites*** are:

(1) ***Hypnomys***, (Rodentia, a giant dormouse) **Upper Pliocene/ Holocene**

Species **(a)** ***waldrenii*** (Reumer 1980) **5-6 m.y.a**
(b) (unpublished)
(c) ***morpheus*** (Bate, 1908) **1 m.y.a. to circa 2200 b.c.**

(2) *Nesiotites, (Insectivora, shrew)* Upper Pliocene/Holocene

Species (a) (Reumer 1980) **5-6 m.y.a**
(b) (unpublished)
(c) ***hildago*** (Bate 1909) **1 m.y.a. until circa 2200 b.c.**

There are divers other non-mammalian species, consisting of birds, reptiles and invertebrates, accompanying the mammalian forms which have been extracted in the hundreds by the use of the hydraulic system from the Muleta deposit (**Waldren 1982: Appendix 1B**). All of these await study by experts having been separated and broadly classified in a preliminary way. The collections of these various microfauna species from Muleta are stored in the **Deya Archaeological Museum and Research Centre, Deya, Mallorca**. Future study of these materials will no doubt cast new light on the question of their role in the chronology and environment of the Upper Pliocene, the Pleistocene and Holocene along with identifying some new species; a new species of frog, ***Baleafrensis muletensis*** (**Adrover 1977**) has recently been identified from the Muleta deposit, shortly after the termination of the hydraulic soil separation campaigns there.

In summary, it is the relative chronology of the fossil record of earlier geological periods and their materials which forms the main evidence supporting a long Presettlement Period of nearly 8-6 million years. However, as will be seen shortly, more accurate chronometric chronological data and materials are available from the Muleta deposit which covers a large part of the Pleistocene and all of the Holocene epochs.

Concerning the Chronometric Data for the Presettlement Period

The chronometric data available for this period consists of 26 dates: 3 of which are on geological materials, 4 of which are racemisation of aspartic acid dates on bone and 8 Uranium Thorium dating of bone and 11 radiocarbon dates on bone; the bone in the case of **RAA, UT** and **14C** being that of ***Myotragus balearicus***. The 26 readings on organic material were the results of a collaborated effort of three American and one British scientific institutions engaged in cross correlating the materials from the Muleta levels using radiocarbon analysis, racemisation of aspartic acid analysis and Uranium Thorium dating (**Part I**). The radiocarbon results were run at the radiocarbon laboratories at the Smithsonian Institute, United States National Museum and the University of California at Los Angeles and the RAA dates processed at the Scripps Institute of Oceanography, University of California at La Jolla. The Uranium Thorium datings were carried out at Harwell Laboratories and University of Oxford. The palaemagnetic dating was conducted by the University of Yale, Peabody Museum and the calcite column core dating at the Quaternary Research Laboratory of the University of Washington at Seattle.

Combined, these chronometric datings form a remarkable sequence which unfortunately has no local comparative example in the Balearic site, as others simply do not exist for the moment; although recently well stratified organic materials and other inorganic materials from a cave adjacent to the Muleta cave producing what is felt to be contemporary deposits are under excavation. Therefore, the group of dates in Inventory have to be treated for the present as a unique series in cave deposit dating. Nevertheless, they do date the various stages of development of the deposit and the materials found in the various levels. From the stratigraphical descriptions found in Part I and in the author's thesis, they follow a natural sequence, considering the complexity of cave deposits in general, and there is little reason that we should question the accuracy of the readings. In fact, as mentioned ealier in this text, the correlation between the results of two different laboratories in the 14C analyses results, as well as the use of three different methods (**RAA, UT and 14C**) seems to be remarkably good.

II

THE EARLY SETTLEMENT PERIOD

<<<<<<<<<<<<<<<<<<<<<<<<<<>>>>>>>>>>>>>>>>>>>>>>>>>>

(circa 5000 b.c. to circa 3000 b.c.)

Recent discoveries in the Eastern Mediterranean (**e.g. the Franchthi Cave on the southeastern coast of Greece**) (**Jacobsen 1976**) illustrate open sea navigation as early as the Seventh Millennium b.c. where in the case of this site obsidian from the Island of Melos, over 150 kms away, was present in the cave, along with tuna fish bones showing deep sea fishing abilities. On the basis of such evidence, a Fourth and Fifth Millennia b.c. origin for the inhabitants of the ESP in Muleta and Matge somes as no surprise. It also permits us to formulate ideas as to the means and route of their arrival in the Balearic Islands.

The geographic situation of the Balearics is such that the islands form a series of '**stepping-stones**' reaching out into the Western Mediterranean, only 94 kilometers from the Valencian Peninsula, from which landfall navigation has always been favourably possible. In the author's opinion, it seems reasonable that the early Balearic settlers of the Early Settlement Period originated from the Iberian Mainland; either as a result of accidental landings (**e.g. coastal fishermen**) or directed and motivated migration. Such movements may well have taken place periodically, not only throughout the Early Settlement Period, but during the subsequent Pretalayotic Period as well.

The lack of domesticated animal species evidence in the Early Settlement Period contexts of both the Muleta and Matge deposits, as well as their 'aceramic' status until circa 2800 b.c. (the hearth debris of the lower levels of both sites contained only kitchen bone debris of ***M. balearicus***) leads the author to reason that only accidental landings may have taken place, rather than motivated migration in the early stage of the ESP. Actually, it is not until the Neolithic Early Ceramic Phase (**NECP**) of the Pretalayotic Period, circa 3000 b.c., that there is direct evidence of motivated migration having taken place. This is demonstrated at Matge and Muertos Gallard in the beginning of the Pretalayotic Period by the appearance of pottery technology circa 2800 bc, found in these lower hearth levels; pottery forms comparable to Iberian and European Late Neolithic types. These lower hearth levels also mark at Matge the first appearance of domesticated animals (**goat, pig and small cattle**) alongside ***M. balearicus*** remains. As radiocarbon dating shows, it is not until the higher level of the NECP (**Stratum 26**) that the final disappearance of the ***M. balearicus*** occurs circa 2150 b.c. (**Burleigh and Clutton-Brock 1980**).

The proposal for the formal introduction to the Balearic prehistoric sequence of an Early Settlement Period is based on the discovery of evidence of the earliest human settlement as yet found on the islands, originating at the two primary research sites of Muleta and Matge, and toward the end of the Period the secondary site of Muertos Gallard.

The three sites are geographically separated by about 30 kilometers and yet have produced similar chronological evidence for man in the Balearics. However the latter station of Muertos Gallard is, while on the one hand, not linked with Muleta and Matge at the inferior end of the Settlement Period, circa 5000 b.c. and but at the end of the period at Matge, circa 3000 b.c. Hence, Muertos Gallard and the comparable levels at Matge form an interface between this period and that of the subsequent Pretalayotic one; an aspect further examined below.

The combined evidence found in these three sites offers a variety of different types of documentation relating to man's first settlement of the islands and his activities there during a span of time of some 2000 years. These discoveries are of considerable importance, since at one stroke they double the formerly assumed duration of the entire cultural prehistoric record from the start of the Pretalayotic Period to the end of the Post Talayotic Period.

The history of this discovery began in 1962 at Muleta and started with the discovery of the first human skeletal remains directly associated with ***M. balearicus***, dated circa 3985 ±109 b.c. It continued with the further findings in 1968 at Matge, where that site produced abundant evidence of man and ***Myotragus*** circa 4730 b.c.

Between the two stations a considerable amount of physical evidence has been amassed, all of which is treated in detail in the appropriate sections of the author' doctoral thesis (**Waldren 1982**) However, an over view of the types of supporting evidence is necesary, if only in summary form. The evidence summarised below is certainly of sufficient quality and variety to firmly describe the period. However before any summary is undertaken a bit more background is necessary.

In the interim between the discovery of the first evidence at Muleta and the subsequent Matge findings in 1968, the materials from Muleta stood as the sole evidence for Balearic early human settlement in or around the Fourth Millennium b.c. It was to remain so until the author's discovery of the rock shelter of Son Matge. The original Muleta materials, being unique as evidence for much earlier habitation of the islands than assumed, were in need of substantiation by similar finds at another site, and Matge was to more than fill that role. The discovery of human remains of proven antiquity at Muleta was well and good, but it did not in itself offer much clue to the reason for man's presence there or, for that matter the extent of the human presence on Mallorca.

The discovery of a second source of evidence of early man at Matge, not too far distant from the Muleta site, and with very similar chronology, not only establishes without a doubt a much earlier date for man's presence in the islands, but also gives us a rough idea of distribution and, in a sense, preferred locales of early human settlement in the mountain regions of Mallorca. It also establishes a very close relationship with the ***Myotragus balearicus***, whose preferential habitat was also the mountains.

As we will see, whereas the Muleta cave evidence gives us the first indications of such a relationship in the stratigraphical contexts of the Presettlement Period at Muleta, the varied evidence of intensified human activities at Matge <u>establishes</u> as well as <u>defines</u> an Early Settlement Period. We can also begin to examine and begin to fill in details regarding the extent and duration of these relationships suggested in the combined evidence at Muleta and Matge.

While the precise source area for the first settlers of the Balearics at present can only be guessed as coming from some part of the Iberian Mainland- probably a coastal region and, hence, likely coastal fishermen and families- it is too early to speculate on population during this Early Settlement Period, although it does appear probable from the distribution at present demonstrated at Muleta and Matge that the number was not too great, most likely confined as it was to geographically dispersed small groups. Equally, either site area would have had its advantages, but from different standpoints, as one is close to the sea and the other located in the mouth of a narrow gap at the entrance to a mountain valley near the plains. One explanation is the fact that evidence so far has emerged from cave and shelter situations where one might expect good preservation of the evidence; whereas such evidence from the plains might also be expected to be subject to a number of destructive activities, such as soil erosion and agricultural pursuits; as we have pointed out earlier. Another explanation may well involve the fact that the major indigenous fauna, ***Myotragus***, as an essential subsistence requirement, was more prevalent in the mountainous zones of the islands.

The Evidence and Chronometric Data for the Early Settlement Period

<<<<<<<<<<<<<<<<<<<<<<<<<<<<<<<<<<<<<<<<<<<<<<<<<<<<<<>>

Fortunately, there is ample physical evidence from the Early Settlement Period from which we can draw sufficient absolute dating to give us a good idea of the extent and duration of the period. This is possible with the results of 7 radiocarbon dates: 2 from Muleta and 4 from Matge and 1 from Muertos Gallard. There are also another two aberrant 14C dates from Muleta, which were the first two analyses to be conducted on the human and ***Myotragus*** remains. The reasons for this were the size of the test samples first submitted (**Waldren 1982; Appendix 1a**) which are discussed in that appendix; however these do not apply here.

Although mention of the evidence from Muertos gallard is included here for reasons of delineation, at the same time, the author proposes a firm distinction between the near end of the period and the following Neolithic Early Ceramic Phase (**NECP**) of the Pretalayotic Period, circa 3000 b.c. This could be modified at a later date if future research proves it necessary. As we shall see, the distinction between these two periods in their interface stage are no doubt transitional; a recognised probability as such delineations are between all the period and phases forming the framework.

Once again, this series of dates provided by radiocarbon age determination from these research stations have absolutely no parallels or correlations in the Balearics and, so are unique until comparative materials and chronometric dating are eventually discovered. The list of available dates for the Early Settlement Period are listed in the tables provided in Part V.

Two of the sites of Muleta and Matge, as we have seen, have produced exceptionally '**deep**' stratigraphies with extensive radiocarbon sequences, which together form a veritable '**yardstick**' for Balearic prehistory (**Figure 1**).

In each of the site locales, the physical and chronometric evidence have been linked and involved with a human subsistence-economy that is significantly based on the major prehistoric fauna, the ***Myotragus balearicus***, Bate 1909 (**Waldren 1982**). Details of the evidence consists of the following inventory:

(1) Human remains directly associated with ***Myotragus balearicus*** from the cave of Muleta, dated by radiocarbon analysis at (**3985 b.c.**)±109 years (**KBN-640d**) (**SM 54**) (**Stratum 7**) (**Table 5**).

(2) Coprolite levels, butchered bones and trimmed horn cores of ***M. balearicus*** showing tool marks made in dismembering the animals appeared from inferior levels of an animal corral context in the rock shelter of Matge (**Stratum** 35), radiocarbon dated at (**4730 b.c.**)±120 years. (**QL-29**) (**ABSM** 55) (**Table 13**).

(3) Coprolite levels of ***M.balearicus*** (Stratum 34) at Matge, radiocarbon dated at **3870** ±360 years (**CSIC-176**)(**ABSM** 53).

(4) Charcoal from a hearth zone (**Stratum** 33) at Matge, radiocarbon dated at (**3800 b.c.**)±115 years (**I-5516**)(**ABSM** 52).

(5) Charcoal associated with ***M. balearicus***, domesticated animal remains (goat, pig and small cattle) and Neolithic pottery forms (**Stratum 28**) at Matge, dated by radiocarbon analysis at (**2700 b.c.**)±120 years (**QL-988**) (**ABSM 51**). This latter evidence forms the lower interface for Matge's Pretalayotic stratigraphic and chronometric sequence and the upper interface of the Early Settlement Period. Similar dates received recently of (**2810 b.c.**) ±50 years(**BM-1994**) (**AMG 51a**) are also available from the Pretalayotic levels at the rock shelter of Muertos Gallard for the earliest pottery horizons there.

(6) Cultural artefact evidence of a very minor category (**11 bone needles and two small flint flakes**) representing human activity in the Muleta deposit for the Early Settlement Period (**ESP**) levels, along with an equally unimpressive inventory of artefacts (**1 flint nucleus and a number of flint flakes**) from Matge, demonstrating an ESP age, is found in these stratigraphies. It is the exceptionally rich and diverse (**see above**) fire zones and animal bone evidence in lower ESP contexts of both sites that is important. It is mainly due to this lack of significant conventional artefact evidence in these lower levels and early settlement contexts at both sites that the author separates the Early Settlement Period from that of the artefact rich Pretalayotic Period in his chronological scheme for Balearic prehistory. It may be necessary in the near future to alter the interfaces at these levels when more artefact evidence becomes available, as it seems to be cecesary to do with the interface of the Talayotic and Pretalayotic Periods.

(7) A series of four palynological analyses (**Valdren 1966**) (**Table 10**) exists for the Presettlement-Early Settlement Period's contexts of the Muleta deposit, circa 12,000 b.c. to circa 4000 b.c. These delineate the Early Settlement Period at Muleta from that of the Presettlement Period; showing in the pollen rain chart a significant change in the surrounding flora and some indication of cereal crops circa 5000 bc.

Certain Ecological Conditions and their Chronology

Despite the rather limited nature of the information, there is some paleo-ecological data in the form of palynological analyses available for the Presettlement and Early Settlement Periods from the Muleta site

(**Table 10**). Unfortunately their interpretations can be commented on only in the broadest and cursory of terms by examination of the pollen rain chart, where a significant change in the flora coverage in the vicinity of the Muleta cave seems to have occurred and which indicates some degree of agricultural activities.

From the radiocarbon dates of the different levels at Muleta, a considerable ecological change appears to have occurred after circa 5000 b.c.; a change which also coincides with the earliest 14C dates of circa 4730 b.c. at Matge for presence of man. Although many more analyses are needed before we have a clearer picture of the flora of the time emerges, the inference drawn from pollen samples 2 and 3 makes it possible to postulate that agriculture to some degree at least took place in the Muleta area circa 5000 b.c., as there are cereal pollen grains present in the pollen rain chart. Otherwise, the area appears to have been covered by grasses, probably sedges. The presence of grass florets in the archaeological levels of Muleta would indicate that grass may have been brought into the cave, circa 5000 b.c. to circa 4000 b.c. (**Gottesfeld, Martin and Valdren 1981**).

As we will presently see, the Matge rock shelter also has given us other ecological information and evidence, both directly and indirectly, concerning man's early relationship with ***Myotragus balearicus***. The extent and duration of this apparent symbiotic relationship is discussed in the Pretalayotic Period section of this paper. There the present known extent and duration of that relationship will be examined, along with other aspects of the island's first domesticated animal species and artefacts; and in so doing, giving us a clearer and a somewhat more precise idea of the sort of activities and economy present during both the Early Settlement and Pretalayotic Periods, regarding man's use of ***Myotragus***.

In all events, their appears to have been some floral changes from Presettlement Period conditions, circa 12,500 b.c. to Early Settlement Period conditions, circa 5000 b.c. as found in both the Muleta and Matge stratigraphies and which indicates agricultural activities as well as trials at management of **Myotragus balearicus** between circa 5000 b.c. and circa 3000 b.c.

Examination of the Early Settlement Period Evidence

<<<<<<<<<<<<<<<<<<<<<<<<<<<<<<<<<<<<<<<<<<<<<<<<<<<<>>

An inventory of the Early Settlement Period evidence is listed in Tables 4 and 5. The human skeletal materials (**Table 11**), a part of which has been dated by radiocarbon method at 5935 b.p. (**3985 b.c.**) ± 109 years (**SM 54**), represents the remains of three and possibly four individuals found in various sectors of the Muleta deposit (**Stratum 7**). Carbon 14 dating of ***M. balearicus*** bone from 175 cms. (**'0' sector**) directly associated with some of the human remains gave a reading of 7135 b.p. (**5185 b.c.**) ± 80 years (**SM 56**).

The remaining inventory of osteological evidence for the ESP , originates from the Matge rock shelter and consists of unburnt and burnt bones of ***Myotragus*** from the corral levels and succession of hearths found in Matge's East Enclosure (**Figures 4 and 5**). Here, we have a number of radiocarbon dates showing early settlement in the rock shelter, beginning with butchered bones and '**V**' trimmed horns of **Myotragus** found in coprolite bed no.2 (**Stratum 35**) of the corral and which have given us a date of 6680 b.p. (**4730 b.c.**) ± 120 years (**ABSM 55**). The coprolites of ***Myotragus*** in another radiocarbon dating have given us a reading of 5820 b.p. (**3870 b.c.**)± 360 years (**ABSM 53**) for coprolite bed no. 1 (**Stratum 34 - 100 cm. above coprolite bed no. 2**), while man's first hearth fires in the East Enclosure (**Stratum 33**) have given charcoal dates of 5750 b.p. (**3800 b.c.**) ±115 years (**ABSM 52**).

Another important date from charcoal which has to be included in this sequence of the Early Settlement Period, for the reasons stated below, comes from Hearth 5 (**Stratum 28**), which produced not only ***Myotragus*** kitchen debris, but remains of domesticated goat, pig and small cattle, as well as the site's first pottery evidence in the form of Late Neolithic types. Although this level actually belongs to the subsequent Pretalayotic stratigraphic sequence (**Early Neolithic Ceramic Phase**), its radiocarbon date of 4650 b.p. (**2700 b.c.**) ±120 years (**QL-988**) also acts as the upper interface of the Early Settlement Period's stratigraphic sequence. At the same time, it is also a phase of transition, both in the stratigraphical sequence as well as the chronological one; marking as it does the beginning of the levels containing the site's earliest artefact evidence and accompanyng radiocarbon documentation.

Remarks on the Chronometric Dating of the Early Settlement Period

While the present list of absolute dates and materials, with its lower interface of circa 6620 b.c. and upper interface of circa 2700 b.c., the chronometric sequence of seven dates (**Table 9**) is only a beginning, giving us a preliminary or working framework of absolute dates for the Early Settlement Period, undoubtedly future investigations will amplify our knowledge of the period, and eventually it may be necessary to incorporate the ESP into the Pretalayotic Period as a phase, rather than a separate epoch. In all events, it is certain that some adjustment will have to be made, sooner or later, in the chronological scheme as more evidence becomes available, and the extent and nature of man's activities during this early period of settlement becomes better defined.

It seems reasonable to expect that future palynological evidence from the deep stratigraphies of caves and rock shelters, where strict stratigraphic controls have been maintained and extensive chronometric dating techniques have been employed, should be forthcoming. Furthermore, as discoveries so far has shown, it seems more than probable that future evidence of early man in the Balearics will come from such stratigraphies in caves and rock shelters and will be also associated with the endemic species, ***Myotragus balearicus***, where there was a basis of a subsistence economy founded on animal management or hunting.

III

The Pretalayotic Period

circa 3000 b.c. to 1400 b.c.

The Pretalayotic Period was first defined by Colominas and other early investigators as the **Culture of the Caves**, and this has remained the generally accepted definition. It is acceptable to the author as a basically sound description, but recent knowledge shows it to be somewhat over-simplified. Therefore, for the purpose of this current research a slightly more elaborate description of the period will be used. A large quantity of new information is now available concerning the chronology of the period, and its nature and scope are better understood, following intensified excavation and research.

Whereas past investigations were mainly limited to the classification and interpretation of artefacts, many of them found out of context, we now have the first results of long term chronological surveys, carefully studied archaeologically and environmental sequences from sites containing deep stratigraphies, the discovery of new groups of artefacts ***in situ*** , with associated material, and other information which has generally enlarged the potential of our understanding of the Pretalayotic problems. Just as the addition of the Presettlement and Early Settlement Periods have clarified the first stages of Balearic prehistory, so the Pretalayotic Period now stands better defined, and can usefully be subdivided into phases.

The new information on which these subdivisions are based is the result of extensive and carefully planned investigations using as many new disciplines as possible for interpretation, for the first time addressing local ecological problems alongside the purely archaeological ones. In the Balearics, as elsewhere, the priceless opportunity to study long stratigraphies with excellent cultural sequences over the last decade and a half has been one of the basic reasons for the general renewal of interest in insular prehistoric problems, and developments in various techniques for studying palaeoenvironments have kept pace with the new discoveries. It is hardly surprising that the concept of the Pretalayotic Period has generally changed and taken on new meaning to researchers- the author included- as a result of the new findings.

The Pretalayotic Period, as the term is used in the present contexts, is most strikingly separated from the preceding Early Settlement Period (**ESP**) by the first appearance of pottery in the Matge stratigraphy. The transition from the Pretalayotic Period to the Talayotic, on the other hand, is best marked by the establishment of the first permanent **open-air settlements**, circa 2000 b.c., and construction of the first megalithic **Talayot** buildings, circa 1300 b.c. to circa 1200 b.c..

For the significance of these events to be clear, and before we can examine the various phases of the Pretalayotic Period, some recapitulation of features of the ESP and a certain amount of preliminary information on the Talayotic Period will have to be made. Also, the reader should bear in mind that there is considerable chronological and other overlap from each period to the next, rather than really abrupt and clear breaks between them.

In the case of the overlap between the Early settlement Period and the Pretalayotic Period, it is not the appearance of pottery for the first time alone which is used by the author to define the two periods, but also the first appearance of domesticated animals in the Muleta and Matge stratigraphy, though, emphasising the overlap, we should particularly note the late survival of ***Myotragus balearicus*** in the Matge stratigraphy, to circa 2700 b.c., and by recent indications about 2200 b.c. (**Burleigh and Clutton-Brock 1980**).

The overlap between the Pretalayotic and the Talayotic Periods is in turn defined not only by the open settlement and architectural activity, but by two other factors (**a**) the common (**as opposed to occasional**) use of bronze, including its frequent occurrence amongst burial goods, and (**b**) the abandonment of the Matge site as a habitational area and its conversion into a burial ground, circa 1400 b.c., a function which is retained throughout the Talayotic and Post Talayotic Periods.

To stress the essential continuity again, we may note that there is now a fair body of evidence for the presence of some copper working evidence, circa 2000 b.c., and occasional bronze metal objects in the Balearic Islands as early as 1800 b.c., though it is only recently that conclusive proof has been obtained that actual bronze metal working (**as opposed to the importing of ready made artefacts**), took place circa 1800 b.c. to 1700 b.c., (**Waldren 1979**). However, mechanism for the initial introduction of copper and bronze into the indigenous material culture of the Pretalayotic population are still not well understood, and the author tentatively considers the interfaces at the beginning and the end of the Pretalayotic Period to be **transitional phases** liable to rethinking at most any time, though he would guess that any change in them should affect their duration and chronological definition, not their nature. Some flexibility in the reader's thinking is accordingly necessary during the account in this chapter of the phases of the Pretalayotic Period as defined in this research. These phases have been established on the basis of the research site stratigraphies with reference both to the archaeological material in the relevant levels (**see Part II**) and the excellent series of chronometric dates (**Part V**). The phases are three in number as follows:

(1) **The Neolithic Early Ceramic Phase (NECP)**
(2) **The Early Beaker Phase (EBP)**
(3) **The Late Beaker Phase (LBP)**

A

The Neolithic Early Ceramic Phase
circa 3000 b.c to 2000 b.c.

<<<<<<<<<<<<<<<<<<<<<<<<<<<<>>>>>>>>>>>>>>>>>>>>>>>>>>>>>>

This phase of the Pretalayotic Period is a recent addition to the Balearic chronological framework. The evidence supporting it is of three different kinds and can be described as the following: (**a**) the first appearance of pottery in the Matge stratigraphy of Neolithic Mainland typology (**Valdren 1982**), associated with these hearth zones of (**b**) domesticated animals such as goat, pig and a small species of cattle and (**c**) the continued survival of ***Myotragus balearicus*** with these domesticated species in the hearth zones of this phase. There are also comparable pottery evidence from the Muertos Gallard rock shelter stratigraphy. All three of these events occur at the lower interface of the phase, circa 3000 b.c.

These materials and their in Matge and Muertos Gallard stratigraphical contexts have been dated by radiocarbon method, so that there seems little reason to doubt their chronological age or the contexts from which they originate. Stratigraphically, these contexts form a part of a very extensive series of hearths and fire zones which form a succession and sequence of levels covering the whole of the Pretalayotic Period as well as the Early settlement Period contexts (**see Part II, this edition and Valdren 1982: Vol. III, Plates 23:1 and 24:1**). Their chronology covers a period of about 1000 years for this phase and is supported by three radiocarbon dates (see **appropriate section at the end of the period's**

discussion), one close to each of the two interfaces delineating the preceeding ESP and the following NECP of the Pretalayotic; the latter marking the latest survival date for the ***Myotragus***, circa 2300 b.c. Muertos Gallard further supports the ESP/ NECP transitional period with one radiocarbon date.

B

The Early Beaker Phase

circa 2000 b.c. to 1700 b.c.

<<<<<<<<<<<<<<<<<>>>>>>>>>>>>>>>>>

The question of Bell Beaker contact with the Balearic Islands began its debate some fifty years ago, following the discovery of several reconstructable fragments of typical Beaker pottery in Cova dels Bous in Felanitx (**Valdren 1986**). This was a unique find, but surprisingly little importance was attributed to it. Indeed, some authorities regarded the evidence as faked or 'planted'. Fortunately the sherds were recorded and photographed (**Castillo 1928**) before their disappearance. The find was all but forgotten for over thirty years when in 1960 Mallorcan excavations again began to produce material evidence which clearly showed a Beaker presence in the Balearics. Since then, finds have been made with increasing frequency (**Valdren 1970, 1980 and 1984**).

There can be no doubt concerning the authenticity of the new examples, and ample evidence now exists for a serious study of the materials which would have been impossible on the basis of the scanty earlier evidence. In fact, there have already been a few published inventories of the recent discoveries which show the range and quality of the material, and also document the items in chronological order (**Valdren 1970 and Camps Cantarelles 1972**) (**Valdren 1984a and 1986a**)(**Diagram 1**).

When the continental and other insular distribution of Beaker influence and materials are taken into view, it is not at all surprising that Beaker trade activities should have extended to the Balearic Islands. Nevertheless, it is only in the last decade that serious consideration has been given to the problems, motives and origin of Beaker trade and contacts (**Valdren et al. 1986**). The specific problems of Beaker cultural contact in the Balearics are extremely important to the correct understanding of the chronology, origin development and significance of the local Pretalayotic and Talayotic Periods, and their various subdivision.

A programme of research into such matters was started by the author in 1965, aiming first to establish the absolute range of the local Beaker material and to determine its stratigraphic contexts in sites yielding good archaeological sequences. The results were very successful in that they provided a first nucleus of absolute dates, and offered various starting points for more extensive investigations. Even as early as 1965, they clearly showed that Balearic Beaker chronology was compatible with continental dates and that the local material was well within the known Beaker range (**Valdren 1979**). In fact here in 1986, a culmination of much of the past and recent research has been summed up in an **International Western Mediterranean Bell Beaker Conference (Valdren and Kennard in preparation).**

The results of this research programme have been varied. (**a**) They have completely undocumented discoveries prior to 1960. (**b**) They have emphasised the value of such a systematic study and shown the need for a continuing study and the site conditions in which the Beaker evidence is usually found. (**c**) They have made possible further subdivision of the Pretalayotic Period on a sound basis of absolute dating, and (**d**) they have given us a very good idea of the immediate origin of the Beaker influences which reached the Balearics, early navigation and commerce and even, perhaps, some insight into the full Talayotic culture itself. As a side benefit, they have also stimulated new interest amongst continental scholars in an island area which had not previously been regarded as lying within the aspects of Continental Beaker distribution (**Valdren 1986**).

The bulk of the new information now available enables us to attempt more definitive conclusions and interpretations regarding some of the stratified sites and some individual Beaker artefacts (**Valdren 1984**). For example, it is notable that except in rare cases the Balearic Beaker artefacts are usually associated with occupation horizons or burials in caves and shelters. Only very recently have Beaker wares been encountered in association with megalithic buildings; such instances have been of great potential significance, but until recently have offered poor stratigraphical conditions: that is to say Beaker sherds have merely been found either directly on or close to bedrock on which the megalithic buildings were constructed. The author is familiar with at least four new stations- two of which are his own- (**see Part II**) where this occurred. In each instance, the Beaker evidence has been associated with relatively early forms of architecture. These sites are Cotxera (**Rosselló-Bordoy 1973**) and Son Ferrandell-Oleza (**Valdren in preparation**) on Mallorca and Torralba den Salort on Menorca (**Fernandez-Miranda and Valdren 1980**) and Ca Na Costa on the island of Formentera (**Fernandez, Plantalamor and Topp 1977**).

The discovery of megalithic buildings like Ca Na Costa and the others where Beaker wares are associated in this way is an extremely new development. Our chronological survey by radiocarbon has accordingly been extended to the island of Menorca where the samples for dating have been recently collected from the interior of a very old building adjacent to the Taula of Torralba den Salort. These samples were Pretalayotic pottery found immediately below the foundation wall of the building. Any development of this new information will have to appear in a future edition, but the association of the Pretalayotic pottery promised that the new dates will be of vital importance in dating some of the earliest Balearic open-air settlements, and it will be interesting to see how compatible these dates will be with those from the deep stratigraphies of shelters and caves.

It should be emphasised that the Balearic megalithic Talayotic sites show different construction periods which may vary very widely in age. However, up until now, our survey of dates does not clearly demonstrate an early date for Talayotic construction, despite the fact that the various dates obtained from the research sites more than amply cover the chronological period evolved. The oldest relevant date we have so far is from Ferrandell-Oleza Younger Settlement, circa 1200 b.c. to circa 1000 b.c., with an lower interface of 1400 b.c., and this latter interface is based on the abandonment of the Pretalayotic living areas at Matge for the purpose of burial in Talayotic times, circa 1250 b.c., at which time the population must have moved to permanent open-air habitation. Elsewhere, the oldest date directly associated with a megalithic structure with an open-air Talayotic settlement is at Pula, where a date of circa 1310 b.c. (**Rosselló-Bordoy 1978**) has been established.

C

The Late Beaker Phase

(circa 1700 b.c. to circa 1400 b.c.)

The evidence supporting the Beaker phases in the Balearics prehistoric chronological framework can be summed up as follows: (**a**) the present-day inventory of over one thousand sherds of BP and LBP pottery that has emerged from Balearic sites over the last 18 years, mainly from Mallorcan sites, but recently on the other islands of the group, undoubtedly warrants the recognition of a substantial Beaker phase within the Pretalayotic Period. It is immediately clear from the material that it spans a long period of time and demands subdivision. (**b**) The numerous sites producing the pottery evidence over the years have undergone extensive chronological dating specific to the levels from which the evidence originates. The dates fall naturally into two groups, each group covering about 300 years, circa 2000 b.c. to circa 1700 b.c. and circa 1700 b.c. to circa 1400 b.c. (**c**) The dates from the Balearic sites for Beaker material are in accord with each other locally and at the same time they closely compare with dates for similar artefact types found in Continental excavations. Various divisions of the Beaker period on the continent have long been recognised. (**d**) The arrival of the first Beaker influences in the Balearics signal a new development from that of the introduction of a new pottery technology (or as will be discussed, possibly the organised importing of pottery trade goods), but also (**e**) the introduction of metal working and a new interest in the construction of open-air settlements. These things need to be underlined by recognition of a specific phase; it is not sufficient to leave them as part of an undivided Pretalayotic Period. (**f**) It will be shown that the initial stage of the Balearic influence, which lasted about 300 years, was followed by a late Beaker phase that lasted until about 1400 b.c. and by recent evidence later, circa 1300 b.c. to 1200 b.c. This second stage is characterised by changes in pottery techniques, including the way in which the decorative incisions were produced and the patterns made out of them.

The stratigraphical contexts of these two phases are well defined in that the clear technological distinction between the EBP wares and those of the LBP is confirmed both in the stratigraphical successions at the leading sites and also in the chronometric dates obtained. This state of affairs has given birth to two difference terminologies of these wares. They are referred to in local literature as '**type A**' and '**type B**'; however, the present author prefers the clear cut terminology of EBP and LBP wares. The use of the form nomenclature is based on a local general tendency not to accept the Balearic Beaker ware as such, but to treat them as local Pretalayotic wares which happen to be incised. A main aim in this current research has been to establish the truer definition of these wares.

The author considers the upper interface of this phase, circa 1500 b.c. to circa 1400 b.c. as a transitional time, because it will no doubt be in this chronological area that adjustments will have to be made when alterations in the author's system becomes necessary. It is always difficult to single out exact moments of change or developments between two phases, let alone two periods. At the same time, the author believes that many of the changes from this point in time on in local prehistoric events can be traced to events on the Mainland in that they manifest themselves as Mainland influences. As has been noted, it is during the

EBP that metal working technology arrived in the Balearics as attested to by the appearance of Beaker decorated crucible sherds, along with other articles of Beaker assemblage (**Valdren 1979**). It is during this LBP that bronze becomes in more common circulation; although the author does not consider this the Balearic Bronze Age, as will presently be discussed.

There exists a good series of radiocarbon analyses for both the EBP and LBP which do much to confirm the perimeters of the two phases and the Pretalayotic Period itself, when combined with the NECP dates. Furthermore, these dates will be seen to correlate extremely well with Iberian Mainland dates as well as those in the rest of Europe.

The Chronometric Data for the Pretalayotic Period and Its Phases

Radiocarbon dating for the Pretalayotic Period consists of 15 dates for its 3 phases: 2 for the NECP, 7 for the EBP and 6 for the LBP. The interfaces of these phases are areas where changes may occur in the future, as pointed out above with the upper interface of the LBP and the Early Bronze Age of the Talayotic Period. However, as a stand has to be made in drawing a line to the subdivision, the following table (Table 4) provides the reader with the dates available for the Pretalayotic Period.

Similar to the way that the primary sites of Muleta and Matge together form a prehistoric sequence for the Presettlement and Early Settlement Periods, the two primary sites of Matge and Ferrandell-Oleza and their secondary sites of Muertos Gallard and Marroig combined with their occupational evidence, chronometric dating and general contexts offer us an equally exceptional sequence of prehistoric cultural evidence for the Pretalayotic Period, beginning circa 3000 b.c. to circa 1400 b.c. Details of this evidence can be summed up in the following inventory:

(1) Evidence of pottery technology in the form of Late Neolithic pottery types datable to (**2700 b.c**)± 120 years (**QL-988**)(**ABSM 51**). Similar dates of (**2810 b.c.**) ± 50 years (**BM-1994**)(**AMG 51a**) are also found associated with pottery of the same type in the Rock Shelter of Muertos Gallard.

(2) The same level has given us the earliest evidence of domesticated animal species (goat, pig and small cattle) coexisting with the endemic fauna, **Myotragus balearicus** circa 2700 b.c.; a date forming the lower interface of the Neolithic Early Ceramic Phase of the Pretalayotic Period in Matge.

(3) Recent radiocarbon dating of Stratum 26 of the Matge stratigraphy demonstrates the survival of Myotragus balearicus as late as **(2143 b.c.)**± 398 years **(BM-1408)(ABSM-50a)**.

(4) Radiocarbon dating of charcoal from Stratum 22 of the Pretalayotic stratigraphy with Beaker pottery that has been dated at **(2030 b.c.)**± 170 years **(CSIC-178)(ABSM 49)**; a date forming the upper interface of the NECP and the lower interface of the Early Beaker Phase of the Pretalayotic, circa 2000 b.c.

(5) Subsquent radiocarbon datings from Strata 23-17 associated with Beaker pottery have produced dates of(**2020 b.c**). ± 100 years **(QL-5b)(ABSM 480)**, **(1870 b.c.)**± 120 years **(Y-2359)(ABSM 46)** and 1720 ± 100 b.c. **(QL-24)(ABSM 43)**; dates which form the duration of the Early Beaker Phase of the Pretalayotic Period, circa 2000 b.c. At Muertos Gallard a date of **(1840 b.c.)**± 80 b.c. **(Y-1789)(AMG 45)** associated with Beaker pottery is also available.

(6) Early Beaker Phase incised Beaker sherds from a crucible with copper oxide encrustations on them indicate early metal working activities in the Matge shelter circa 2000 b.c. to circa 1700 b.c.

(7) Radiocarbon dating of charcoal from Strata 17-11 of the Matge Pretalayotic stratigraphy associated with Late Beaker incised wares have given us a sequence of dates **(1620 b.c.)**± 80 years**(CSIC-179)(ABSM 42)**, **(1530 b.c.)**± 80 years **(CSIC-180)**, **(1470 b.c.)**± 100 years **(QL-5a)(ABSM 39)** and 1400 ± 60 b.c **(QL-5)(ABSM 38)**; a series defining the duration of the Late Beaker Phase of the Pretalayotic Period, circa 1700 b.c. to circa 1400 b.c. Son Marroig has also given us Late Beaker dates associated with Beaker ware at **(1520 b.c.)**± 80 years**(Y-1856)(SMRG 40)**

(8) Radiocarbon analysis of charcoal associated with EBP incised wares (Figures 20-23) and the interior of the East Wall of the Pretalayotic Settlement of Ferrandell-Oleza (Figures 16-18) have given a reading of(**2000 b.c.**) ± 65 years **(BM-1843)(SFO-OS 47)**; with subsequent dates of **(1840 b.c.)** ± 90 years **(QL-1636)(SFO-OS 45a)** and **(1770 b.c.)**± 35 b.c.**(BM-1981)(SFO-OS 53b)**, **(1750 b.c.)** ± 30 years **(QL-1592)(SFO-OS 43a)** for other contexts of the Ferrandell-Oleza Old Settlement.

(9) Artefact evidence from the Pretalayotic Settlement of Ferrandell-Oleza consists of large quantities of EBP incised wares and indigenous pottery identical to those known from Mallorca rock shelters and caves, dated from 2000 b.c. to 1700 b.c. The other artefact evidence from the site includes copper and bronze awls and ingots, pyramidal and conic 'V' perforated bone, shell and ivory buttons along with many tabular flint serrated dagger and sickle blades (**Valdren 1984: Figures 27-30**).

(10) The architectural elements of town walls and buldings in the Pretalayotic Settlement of Ferrandell-Oleza give us the first indications of the type of early Eneolithic open-air settlement established prior to the Talayotic Bronze Age; at the same time, giving us some idea of the origin of the early settlers, their economy and socio-religious structures, circa 2000 b.c.

IV

The Talayotic Period

circa 1400 b.c. to circa 800 b.c

The Talayotic Period and the Balearic Bronze Age are synomymous in this hypothesis. The start of the period can be defined by certain events that constitute changes which separate it from the former Pretalayotic Period. (**a**) It is the period when most, if not all, the megalithic or Talayotic architecture of the Balearics were constructed, and it predominantly marks the moment when man abandoned living in the caves and shelters and lived in the open-air settlements of free standing dwellings. (**b**) It is the period in which there first began to be a common used of bronze and extensive metal working as trade and contact with adjacent continental areas. (**c**) During this period, there was a dramatic change from the Pretalayotic times in the typology and technology of the pottery and indeed of the artefact assemblage as a whole. (**d**) Cremation burials replace the Pretalayotic tradition of inhumation and secondary burialsduring this period.

While the facts above separate the Talayotic Period from the former Pretalayotic Period some overlap can be expected between the periods and their phases, as indeed the name 'Pre' Talayotic in part implies, and adjustments will undoubtedly have to be made in the chronological framework as more evidence becomes available. An example of this is recent evidence that actual local metal working occurred in the EBP and LBP of the Pretalayotic Period, circa 1800 b.c to 1700 b.c. This certainly signals the introduction of bronze working into the Balearics, though there is some indication by recent analysis results from samples of the early metal artefacts that a true tin bronze was not in use until 1700 b.c. (**Waldren 1979**). However, in the author's mind, there is still not enough evidence to warrant defining either the EBP or LBP of the Pretalayotic Period as an Early Bronze Age.

Here in 1986, it has been demonstrated that the position taken in 1979 by the author was the correct one concerning this chronological phase, quote:

"If anything, the author considers it (EBP of the Pretalayotic) an Introductory Bronze Age, the limits of which are not yet sufficiently defined within the EBP or LBP. Therefore, he considers a date circa 1400 b.c., or perehaps even later, as a tentative date for the Balearic Bronze Age proper. Undoubtedly this assessment will be the subject of one of the first such adjustments of the chronological framework referred to above; indeed new information is already beginning to emerge at the writing of this edition."

During this interim between 1979 and 1986, the tentative position concerning the transitional phase between the NECP of the Pretalayotic Period and the lower Pretalayotic Period, the evidence does incourage that assessment.

We have also come to see that at the end of the Pretalayotic Period or LBP, a similar transitional period exists between the Early Bronze Age of the Talayotic Period and this late Pretalayotic phase, circa 1300 b.c. to 1200 b.c. circa 800 b.c. Again, another similar transitional phase exists in the dating of the Early Iron Age of the Post Talayotic Period and that of the Late Bronze Age Talayotic, circa 800 b.c. and the limits of the actual assessment of the dividing date between the Talayotic and Post Talayotic Periods is a tentative one which will also probably go through adjustment in the near future; and once more in this recent edition, the author has formulated his chronological framework assessments based on this factor. 'Working' dates are always needed, and it never matters if they are later shown to be incorrect. It is not so long ago that the European Neolithic was held to begin around 2000-3000 B.C.

A

The Early Bronze Age

circa 1400 b.c. to circa 1300 b.c

The evidence to date indicates that the beginning of the Early Bronze Age is circa 1400 b.c, although it may eventually be shown that the phase's beginning is closer to 1300 b.c. to 1200 b.c. The EBA is marked by a series of visible changes which have already been mentioned as indicating the transition from the LBP of the Pretalayotic Period. Amongst these, one of the most striking is the notable change in pottery typology and in the manufacturing techniques such as the clay fabrics (**Valdren 1982: Appendix 3A**). Another visible change between the two periods is the great increase in the number of bronze artefacts, especially those found with burials, which certainly testifies to the increase in availability of the metal and an increasing affluence. A third innovation lies in the radical change of burial custom from earth burials which show a variety and quality of workmanship not present earlier while the individual items showing some correlation with the Continental Bronze Age cultures. A fourth change is the abandonment of the rock shelters and cave sites such as Matge and Muertos Gallard for living, and what appears to be the first impetus in the construction of proper solidly built settlements. The reader should bear in mind the fact that the earliest date we have for an Early Bronze Age level at the base of a sequence at a building site is at Pula, circa 1310 b.c. However, the pottery evidence as well as radiocarbon dating is identical to EBA levels at Matge and Muertos Gallard; the Matge date being circa 1250 b.c. It is necessary to note that Margaret Murray encountered identical pottery forms in her 1938 excavations (**Murray 1938**) in the Naveta of Torreta in Menorca. This gives some idea not only of just how strongly entrenched the traditions of this pottery style became, but also of how widely spread it was geographically.

B

The Middle Bronze Age

circa 1300 b.c. to circa 1000 b.c.

This phase of the Talayotic Bronze Age represents the apogee of the Talayotic culture. The manifestation of this includes extremely rich artefact evidence both in objects of everyday life and in grave goods (illustrated in the appropriate section). This abundance in the burials of the time suggests an unusual affluence for an insular situation where there is nearly none of the necessary raw materials necessary. Furthermore, one need only consider the geographical location of the Matge site to note that this affluence reached even into the mountainous zones of Mallorca. During this phase the settlements such as S'Illot, Son Oms, Ses Paisses and Son Real on Mallorca and Son Catlar, Torre den Gaumes, Torralba den Salort and others on Menorca were clearly very active communities. There is no great change in pottery styles which remain essentially the same. There does seem to be some evidence of architectural renovation in the settlements circa 1000 b.c.

C

The Late Bronze Age

circa 1200 b.c. to circa 800 b.c.

A change in pottery style occurs during this final phase of the Talayotic Period (as illustrated in the appropriate section) as indicated from a pottery kiln of this time at Matge. The first iron artefacts appear in lower levels associated with this kiln, circa 800 b.c. It is according to Rosselló-Bordoy (1973) during this phase that there are indications of the fortification of some of the settlements, which in the author's opinion was probably intended as a def ense against attack from the classical Mediterranean world, although the main direction of cultural influence was still from the Mainland area to the west and to the north of the Balearics.

The Chronometric Data for the Talayotic Period and Its Phases

This Talayotic Period and its phases are supported by an extensive series of radiocarbon readings from a wide variety of sites of Talayotic origin, mostly concerning contexts within the settlements of the period, but also from cemeteries.

It is the transitional stage between the latter LBP of the Pretalayotic and the Early Bronze Age of the Talayotic Period in which recent analysis and artefact findings have brought about new thinking concerning the interfaces of both these periods. The problems of adjustment and re-evaluation of these chronometric interfaces are discussed in fuller detail in the Conclusions and Observations (**Part IV**), but some explanation is also included in the following summary of results.

There are at present 22 analyses results which serve as the interfaces of the two periods: 13 for the EBA, 4 for the MBA and 5 for the LBA. The summary of evidence and these dates can be found below.

(**1**) Of the seven dates available for the Early Bronze Age, only 4 originate from the primary sites (**1260 b.c.**) ± **80** years (**BM-1677**) (**SFO-OS 36a**); (**1250 b.c.**) ± **100** years (**Y-2667**)(**ABSM 36**); (**1145 b.c.**) ± 70 years (**BM-1698**)(**SFO-OS 35a**) and (**1120 b.c.**)± **50** years (**HAR-3490**)(**SFO-OS 34c**). The remainder are from either secondary or tertiary research sites. The three most pertinent in the secondary research sites are from Torralba den Salort on the island of Minorca (**1080 b.c.**) ± **70** years (**QI-1433**)(**TT 34b**); (**1070 b.c.**) ± 60 years (**HAR-?**)(**TT 34a**) and (**1020 b.c.**) ± **70** years (**HAR-2908**)(**TT 33a**). And as we will see most of these can be equally interpreted as Late Pretalayotic in origin. What appears here is a need for many more dates from carefully excavated settlements before real detail to the transitional phase of both period is understood. We may well be dealing with two different coexisting cultures or a much later survival of Pretalayotic elements, or for that matter erroneous interpretation of site levels.

(**2**) In fact, it is in the Middle Bronze Age, where we see the indications that the Talayotic Period is not as early in origin as we have hitherto been led to believe. This is seen in the primary site of Ferrandell-Oleza, Talayot *1, where construction dates are established by radiocarbon sample results from beneath the foundations of both the central pilar and the wall foundations as (**960 b.c.**)± **100** years (**HAR-3413**)(**SFO-YS 30a**). Late abandonment dates for the uppermost Pretalayotic LBP phase at Ferrandell-Oleza Old Settlement, **(1145 b.c.)** ± 70 years **(BM-1698)(SFO-OS 35a)** and **(1120 b.c.)** ± 50 years **(HAR-3490)(SFO-YS 34c)**, also substantiate this by not only showing a much later abandonment of the site but also dating the construction of Talayot *1 of the Younger Settlement. The pottery typological evidence also supports these conclusions. More concerning these aspects are also found in the Conclusions and Pbservations.

(3) For the Late Bronze Age, the interfaces on both ends of the phase are somewhat clearer, although there are still problems of minor interpretation. The best indicator for the upper interface of the Late Bronze Age comes from what are believed to be construction or preconstruction dates for the Taula Sanctuary of Torralba den Salort, where dates of **(890 b.c.)** ± 30 years **(QL-1089)(TT 29a)** from organic materials in bedrock floor crevices inside the Taula and **(880 b.c.)** ± 40 years **(QL-1164)(TT 29)** for charcoal from under the outside wall of the Taula dates its construction or preconstruction. It is considered by the author that these Taula structures, especially Torralba den Salort, were the last and most sophisticate of Talayotic constructions, and from this point forward Talayotic construction begins a decline both in techniques and quality.

The uppermost interface of the Late Bronz Age, is marked by the further introduction of new ideas, technology and even indication of new arrivals. The introduction of Iron is the main delineator as is the change in burial customs etc. We shall consider these points of delineation with the evidence and dates of the Post Talayotic Iron Age reviewed in the following section.

V

The Post Talayotic Period

circa 800 b.c. to circa 100 b.c.

<<<<<<<<<<<<<<<<<<<<>>>>>>>>>>>>>>>>>>>>>

The Post Talayotic Period and the Balearic Iron Age are synonymous in this hypothesis. There are four major cultural changes which define the Post Talayotic Period, differentiating it from the Talayotic Period and its Bronze Age subdivisions. Those major cultural changes are: (**a**) the introduction and common use of iron, (**b**) the distinct change of burial customs, (**c**) the changes in pottery styles and techniques, as well as typological changes in artefacts, including the introduction and use of other metals and materials (e.g. lead and glass) and (**d**) the decline of traditional Talayotic building.

All these changes are undoubtedly interrelated and while, at first glance, they might appear to have occurred simultaneously and very rapidly, deriving from the '**parent**' Talayotic Culture, they were actually the result of gradual change, lasting over a period of a century or so. At the same time, it is the author's opinion that the Post Talayotic Iron Age of the Balearics coincides with the Mainland '**Urnfield**' invasions of Catalona in the northeast, and events which transpired in that region circa 800 b.c. to 700 b.c.; not only on the Mainland and Continental Europe, but also they reflect the influences awakened at this time in the Mediterranean, such as the increased '**classical**' trade and colonisation. These new eastern external influences, along with the acceleration of change in the west, are in the author's mind the forces that brought about the listed changes which distinguish the Post Talayotic Period from the preceding Talayotic one.

The frame work chronologically delineates the Post Talayotic Iron Age and its subdivisions as covering a period of about 700 years and suggests a data circa 780 b.c. for the first introduction of iron, occurring close to or just after the LBA of the Talayotic Period. This important assessment is based on the radiocarbon documented evidence from the Matge stratigraphy, and such a date correlates remarkably well with Mainland dates for the Celtic Catalonia '**Urnfield**' invasions, circa 750 b.c. (**Savory 1968**), in which iron objects occur regularly in burials.

A

The Early Iron Age

circa 800 b.c. to circa 600 b.c.

<<<<<<<<<<<<<<<<<<<>>>>>>>>>>>>>>>>>>>>>

This Early Iron Age phase is a transitional one from the LBA of the Talayotic Period and hence the author uses the date of 800 b.c. for the delineation of the period. While this also marks the approximate date for the introduction of iron, he believes that a date of circa 750 b.c. is more acceptable for the introduction of iron into the Balearics, and that this latter assessment is in more agreement with Continental accepted 14C dates. The evidence, both 14C dating and stratigraphical, at Matge indicates dates of circa 690 b.c. in one sector, (**West Enclosure**) associated with quicklime burials, while the non-quicklime cremation burials of two sectors in the East and West Enclosures gave charcoal 14C readings of circa 780 b.c. and 750 b.c., respectively. Other test samples have been collected during work but have not as yet been processed. Actually, more test results are needed to remove the ambiguity between the LBA and EIA phases; a situation which might be thought somewhat similar for the Hallstatt/La Tene transitional phase in Europe, in some areas at least.

Despite what ambiguity that may exist, the EIA phase of the Post Talayotic Period is marked at Matge and at Muertos Gallard by the first appearance of iron objects in those sites; typical iron objects such as an iron shafted spear, iron socketed spearheads, daggers and antennaed iron short swords, all of Mainland similar types reported there (**Waldren 1982**).

The pottery types of this phase show little change from the LBA wares, other than innovations of the '**imitation**' potter's foot and the appearance of the loop-eye or ribbon handle for the first time on traditional LBA forms. The clay fabric of the EIA wares are also very much the same, being heavily tempered and similar in firing.

The overall artefact assemblage of the EIA is not as strikingly varied as it becomes in the following MIA phase, thus giving the author the impression that this phase of the Post Talayotic Iron Age is a transitional one, highlighted by the introduction of iron and the change of burial customs ;the appearance of quicklime as a burial medium. These two events which are outstanding suggests the possible '**invasion**' of Mainland influences or even people bringing with them new influences and opening up new trade and routes of trade other than the traditional ones; in short the true Iron Age of the Urnfield culture.

Regarding the architectural activities of this time, circa 800 b.c., there is evidence, as outlined earlier, that building activities had lost most of their impetus by this Early Iron phase. For example, recent evidence suggests that the Taula sanctuaries of Menorca were probably the last of the major Talayotic architectural achievements in the Balearic building sequence. This evidence comes from the Talayotic settlement of Torralba den Salort (**Fernandez-Miranda and Valdren 1981, in press**) which has given us important radiocarbon dates for the first time concerning the construction of the Taula sanctuary there. It appears that an older building adjacent to the Taula sanctuary was partly destroyed circa 900 b.c. to enable the construction of the Taula sanctuary. As pointed out in thediscuaaion of the preceding Late Bronze Age contexts, two important radiocarbon dates of circa 890 b.c. and 880 b.c. were taken from two different levels in two widely spaced areas,. one coinciding with the construction level inside the sanctuary and another outside and under the wall of the building; thus dating its construction about 800 b.c. to 980 b.c.; thus supporting this argument.

B

The Middle Iron Age

circa 600 b.c. to circa 400 b.c.

<<<<<<<<<<<<<<<<<<<<<<>>>>>>>>>>>>>>>>>>>>>>

In the MIA of the Post Talayotic Period, there seems to have been a rather prosperous material life style in the settlements, as shown by the wide variety and quality of the iron and bronze goods that are available at least for burial purposes. Toward the end of this phase other classes of items appear for the first time in the artefact assemblages, such as lead and glass. The first important changes appear here in the EIA/MIA wares in the form of well executed loop-eye and ribbon handles (see text).

While quicklime inhumation is the principal mode of burial during all of the Post Talayotic Iron Ages, there are other burial methods which can be described, which may indicate the arrival of still other exotic influences in one form or another; along side the survival of earlier ones. This is demonstrated by the presence of more elaborate individual custom of burial which takes place in especially constructed stone tombs, such as those found in the seaside necropolis of Son Real and Isla del Porros on Mallorca (**Tarradell 1964**) (**also see Conclusions and Observations, Part IV**). Here individual sepulchers were constructed to represent the different forms of Talayotic architecture in miniature- a far cry from the finely built burial navetas of the Talayotic Period; although these miniature tombs are well constructed, they do not to the author constitute megalithic Talayotic building. A 14C date of circa 480 b.c. with a large statistical error of plus or minus 300 years exists for this important site and simply succeed in telling us that it is Iron Age; that and the fact that the graves contain iron artefacts. However, it does not date the individual tombs themselves as the date comes from a communal burial chamber on the nearby island of Isla del Porro, which is believed to be part of the same cemetery. The author is at present working with M. Tarradell of the University of Barcelona in a dating survey by radiocarbon on bone samples from the various types of tombs in the necropolis. The site is probably one of the most important sites of its kind in the Western Mediterranean and its closer study, especially related to an extensive 14C survey, would certainly be informative from the point of view of many aspects. Trace lime has also been found in some of the tombs (**Tarradell 1965**) and the grave goods includes iron artefacts of the same typology as those emerging from Matge and Muertos Gallard, so there is little doubt of their relative chronology.

Also, according to radiocarbon evidence and pottery types, we can attribute to the MIA a few cave burials where wooden coffins have been used in burials. Recent British Museum 14C dates of circa 450 b.c. and circa 400 b.c. date these. In still another instance similar sarcophagi have been burnt in the rock shelter of Son Maimo (**Amoros and Rossello-Bordoy 1973**). The unburnt sarcophagi of Son Boronal, Mallorca (**Lopez in press**), were made of larch and spruce- as identified by Jodrell Laboratory, Botanical Gardens- (**Cutler 1980, written communication**)' the <u>larch</u> tree is not a local tree species, and hence suggests importation.

Although the Carthaginian colony of Ibiza was founded in 654 B.C., classical pottery does not appear in the settlements during this phase until the end of the Middle Iron Age, circa 400 b.c., but some excellent examples of classical bronze seems to have been in circulation (**Rossello Bordoy 1973**; **Fernandez-Miranda 1979**). Fernandez-Miranda has illustrated about half a dozen Carthaginian bronzes in his thesis. Rossello Bordoy recently excavated a small statue of the Egyptian Pharoah Imhotep, Son of Ptah, in the Taula sanctuary of Torre den Gaumes on Menorca, which he dates at 663 b.c. to 525 b.c. (**Rossello Bordoy et al. 1974**). The present author has recently also excavated a bronze bull in a late context, circa 200 b.c., at the Taula of Torralba den Salort, Menorca (**Fernandez-Miranda and Valdren 1981, in press**), although it is believed to belong more to the 4th or 5th century b.c. These important as well as beautiful finds give a good idea of the classical world contacts, best demonstrating the wide sphere of contact of this EIA and MIA period of the Post Talayotic Iron Age.

C

The Late Iron Age

circa 400 b.c. to circa 100 b.c.

This is the last phase of the Balearic prehistoric record considered in this report, despite the continuance of urban life and the traditional customs of the Post Talayotic Iron Ago into the Roman colonisation and the Post Colonisation Period - a subject of separate research. The LIA of the Post Talayotic Period shows the accumulation of all the major changes in the '**parent**' Talayotic Bronze Age outlined and discussed earlier in this chapter.

By 300 b.c., the indigenous pottery assemblage shows marked traces of change in the traditional skills of pottery manufacture. The pottery takes on the quality of a very crude pottery, not unlike the ***poterie grossiere*** of Iron Age Catalonia and southern France. The pottery fabrics are full of organic materials, having been tempered with what appears to be grain which results in a very porous product. The forms themselves are very crude and imitative of classical forms. These '**crude**' wares are found alongside classical imports, such as Attica and Campanian wares along with Punic amphorae, as they are on the Continent. There is a literal saturation of classical wares in the local settlements from this LIA phase forward.

Life in the settlements seems to be accelerated, perhaps aided by a rising population, military ventures and new areas of trade etc.; but at the price of a severe decline in earlier Talayotic tradition. Balearic mercenaries have begun to establish reputations in the classical world in the 4th and 3rd centuries b.c. It is at this point in time that all major building in the Talayotic '**tradition**' has completely ceased, probably partly due to whatever work forces available for large stone building being employed for other ventures dealing with military involvement, trade and the needs of growing population. By 123 B.C., the islands are colonised by the Romans, having acquired them as an outcome of the Punic Wars and in turn establishing the Roman towns of Palma and Pollentia on Mallorca, Mahon and Alayor on Menorca; a situation never accomplished by the Carthaginians. The accounts of early classical chronicles by historians like Diodorus, Strabo and Pliny have left written record of the local Balearic involvement in Military ventures, such as the use of 5000 mecenaries in a single campaign against Sicily.

The Chronometric Data for the Post Talayotic Period and Its Phases

There are 48 radiocarbon age determinations for the Post Talayotic Period and its phases: 17 for the LIA, 20 for the MIA and 11 for the EIA, a listing of which is found in the Inventory. The main reason for such a large number of Iron Age dates in particular is due to the nature of the sites and materials, and secondly the desire on the part of the author to date these late contexts as well as the Talayotic question in general. The concept of a Balearic Bronze Age as well as an Iron Age is a new one and until the present most investigators have been satisfied to consider both periods within the contexts of a number of Talayotic phases without much regard for Continental counterparts; e.g. Rossello Bordoy's Talayotic phases I-IV and Fernandez-Miranda's Talayotic phases I and II. As pointed out earlier in a section above, these chronological frameworks are conservative ones which for the present author are too general, regarding the question of prehistoric events strictly from an insular point of view. While the present author's scheme for these two periods may be a very ambitious one, he feels, like he does about the other periods, that a better assessment is possible. With the very large number of well dated contexts and sites currently available, a more complete chronological scheme, which incorporates mainland nomenclature and parallels, should be possible; even if considerable adjustment periodically should prove necessary within the subdivisions. The list below actually supplies dates at 20 to 30 year intervals for a 700 year period, between circa 80 b.c and 780 b.c. There are also several A.D. readings which are not included in the listing as these are associated with classical coins and pottery of known historic age as well as being outside the limits of the present study, or they are results from superficial levels representing field clearance activities.

The supporting chronometric and other evidence for the Post Talayotic Iron Ages is found in the following listing.

(1) At Matge, the Early Iron Age is represented by the two dates on charcoal for the introduction of iron in the East and West Enclosures of the rock shelter, when it was used as cemetery: **(780 b.c.)** ± 100 years **(QL-7) (ABSM 26)** and **(750 b.c.)** ± 80 years **(QL-11)(ABSM 25)**. Another date, one on carbonate from a quicklime burial sample for the middle of the phase is **(690 b.c.)** ± 100 years **(QL-27)(ABSM 23)**, representing the earliest date at Matge for inhumation in quicklime.

The remainder of the dates for this phase originate from a number of tertiary sites research sites, mainly open-air Talayotic settlements, such as Son Fornes: (750 **b.c.**) ± 170 years (**UGRA-123**) (**FORN 24a**); (**610 b.c.**) ± 85 years (**UGRA-381**) (**FORN 21a**) and two at (**590 b.c.**) ± 80 years and 85 years (**URGA-382 and 124**) (**FORN 19a and 19b**). S'Illot also has an equivalent early date of (**740 b.c.**) ± 60 years (**HU-1718**)(**SILL 24**)

(**2**) The Middle Iron Age is well represented by both a number of settlements and the stratigraphical contexts of Matge. Among the settlement dates are those from the Ferrandell-Oleza Younger Settlement in contexts, especially from the Talayots *1 and *2, dated from (**590 b.c.**) ± 80 years (**HAR--3458**)(**SFO-YS 19c**) and (**510 b.c.**) ± 80 years(**BM- 1842**)(**SFO-YS 15a**) in Talayot *1 and in Talayot *2 at (**450 b.c.**) ± 60 years (**I-4584**)(**SFO-YS 13b**). Matge rock shelter also has a series beginning with (**590 b.c.**) ± 100 years (**QL-4**)(**ABSM 21**); (**590 b.c.**) ± 80 years (**QL-24**)(**ABSM 20**); (**570 b.c.**) ± 80 years (**QL-6**)(**ABSM 19**); (**530 b.c.**) ± to years (**QL-10**)(**ABSM 17**); (**480 b.c.**) ± 230 years (**BM-1842**)(**SFO-YS 15a**); a date with a large statistical error.

There are a good number of dates from an assortment of sites with Middle Iron Age contexts, again mainly from open-air situations (**see inventory listing and Conclusions and Observations**). A few of these are in settlements like Son Fornes; the series beginning at: (**590 b.c.**) ± 80 years (**UGRA-124**)(**FORN 19b**); (**540 b.c.**) ± 130 years (**UGRA-122**)(**FORN 18a**) and (**500 b.c**) ± 190 years (**UGRA-121**)(**FORN 16a**).

(**3**) The Late Iron Age is well represented over the whole range of research sites and from a large spectrum of situations; e.g. burials (**cremations, quicklime inhumations, earth burials and secondary burials and burial in assorted types of tombs, including cremation in wooden sarcophagi etc.**) as well as settlement loci. At Matge dates range from (**340 b.c.**) ± 100 years (**QL-5C**)(**ABSM 12)** to (**120 b.c.**) ± 120 years (**QL-7A**) (**ABSM 2**).

This brings to a close our discussion of the Pentapartite Division of Balearic Prehistory.

IV

OBSERVATIONS, CONCLUSIONS AND BIBLIOGRAPHY

In the following sections, a summary in the form of observations as well as conclusions are presented which outline the various research criteria that have been employed in obtaining the results reported on in the text. As the research itself has spread out over more than two decades since its inception in 1962, it has been periodically necessary to shift the focal point of the research strategy from one stage of interpretation to another. However on the whole the focal point or emphasis have been placed on the sequential order set down in the following outline of research criteria.

In one sense, it has been difficult to reach conclusions in any form other than tentative ones as the results do more than any one thing to speak for themselves; each result being a small conclusion in itself. On the other hand, while the research programme has and still does involve both on-going excavation and continued long range dating results for the excavation's contexts, the purpose of this particular area of investigation has not altered since the formation of the first nucleus of absolute dates compiled in 1962. Therefore, it cannot be concluded that this current reports is in any way a definitive one. On the contrary, it is certain that a periodic up-dating of the present report will prove necessary eventually with the addition of a score or more readings pending from the research sites still under study. It is also possible that the addition of new Balearic site results from radiocarbon method and other dating techniques will eventually also cause the revision and extension of the present edition; hopefully to offer us as successful results as those obtained so far.

While it has been the general rule that all stages of interpretation have been considered throughout, the immediate focal point in the last few years has dealt mainly with the fifth and sixth stages outlined in the research strategy guidelines listed and described below; with the others listed forming a background of activity. The present publication is, therefore, the result of all these research strategy guidelines and interpretative stages and their formal presentation below are one of the observations and conclusions formed regarding the past research in the Balearic Islands as well as being the guidelines for any future work by the author.

Research Guideline Strategy and Stages of Interpretation

The strategy or logical order of approach and the stages of interpretation set down in the present research can be summerised in the following description.

(1) **Chronometrical and Cultural Considerations**: On the simplest of interpretative levels, the objective has been to place particular item artefacts within a certain time frame.

(2) **Contextual Organisation**: On the second interpretative level, the objective has been to delineate and assess the contexts from which the item artefacts have originated.

(3) **Interrelated Contextual Organisation**: On the third interpretative level, the objective has been to evaluate the relationships between the item artefacts and their contexts with regard to other such associated samples.

(4) **Comparative Interrelated Contextual Organisation**: On the fourth interpretative level, once a series of such item artefact and context relationships have been established as a fund or source basis, a comparison and correlation with those of other similar geographically localised situations have been taken into consideration; especially where similar treatment has been carried out or similar results obtained.

(5) **Geographic Comparisons and Correlation**: On the fifth interpretative level, there has been a comparison and correlation of geographically localised situations with those of other geographic situations farther afield.

(6) **Interpretative Tools, Samples and Techniques**: On the sixth and last interpretative level, there has been a multi-varied use of conventional and unconventional materials and techniques of analysis.

The following is a descriptive summation concerning the research methodology in regards to the research strategy guidelines.

Research Methodology

‹‹‹‹‹‹‹‹‹‹‹‹‹‹‹‹‹‹››››››››››››››››››

Interpretative Samples and Techniques: these can be summed up under the following considerations: (1) a use of as many different kinds of materials as possible, (2) a use of as many different methods of analysis as possible and (3) a use of as many individual site contexts as well as different site locales as possible.

(1) In the first consideration, the use of as many different kinds of materials as possible refers to the **interpretative sample**.

(2) In the second consideration, the use of as many different methods of analysis as possible refers to the **interpretative technique**.

(3) In the third consideration and the most varied of the three, the use of as many individual site contexts as well as different site locales as possible refers to the **interpretative problem**.

‹‹‹‹‹‹‹‹››››››››

Interpretative Problem: this aspect takes on a number of perspectives, each different in nature. They are set down in the following main categories along side their pertinent and individual problems:

Geological: This category has considered the problems of geology in terms of **caves**, **rock shelters** and **open-air** sites.

Geographical: This category has considered the geographic situations of **mountain**, **valley** and **plain** sites.

Functional: This category has considered the use of a specific site arena of activity, such as **domestic**, **work zones**, **worship** and **burial**. Each arena of activity has sub-level qualifications of number and type.

Depositional: This category has considered the character of each site's individual stratigraphy in terms of **vertical** and **horizontal** distribution, or **combination** of both.

The research and the results reported on here have demonstrated examples of each of the preceding problems as well as the samples and techniques used for interpretation in the formation of the Pentapartite Division of Balearic Prehistory.

Conclusions

<<<<<<<<>>>>>>>>>>

Apart from being an excellent reservoir for dating techniques and the results obtained from them, we can conclude that the insularity of the Balearic Islands has no doubt been the controlling factor in the dissemination and absorption of the life forms and cultural influences arriving on them from abroad; a mechanism or process of sorts that could take place quite rapidly. At the same time, this insular mechanism or process is seen by the author as being stimulated, in the case of the Balearics, by a succession of major arrivals and series of events which occurred at sporadic intervals throughout the prehistory of the islands. Each of the arrivals in the sequence differed somewhat from the first accidental landings in that the people concerned after the intitial arrivals, brought with them important new technological or cultural contributions to be added to the cumulative indigenous heritage; contributions such as domesticated animals, ceramic technology, metal working and even architectural skills. For the sake of metaphor and not at all an inappropriate one, one might refer to these arrivals as '**cultural waves**' reaching the shores of the Balearics from the Mainland, a terminology adopted by the author. The sequence of these **cultural waves** are outlined below.

However before we review these cultural waves and the directional origins from which they came, consideration must be made related to the longer Presettlement Period, where it would appear that the migration of the earliest ***Myotragus*** species, according to the fossil record, occurred from the northeastern direction of Sardinia and the European Mainland by some means such as **land bridge** or **island hopping**, some 8 to 6 million years ago. This is contrary to the author's hypothesis of later eastward Iberian cultural movements into the Balearics. In the case of the **Myotragus** migration, it appears to have occurred when an already existing migration barrier was present between the large island mass of Menorca-Mallorca and the ancient Island of Ibiza- Iberian Mainland land mass. This is attested to by the complete lack of fossil remains of ***Myotragus*** todate on either Ibiza or the Valencian Mainland connecting regions (**Valdren 1984, Chapter I, Vol. I**).

Radiocarbon dating of the ***Myotragus*** remains from the Muleta deposit and subsequently from the Matge rock shelter have demonstrated the animal's survival all through the Holocene until its extinction directly and indirectly by the hand of man about 2200 b.c., the currently established 14C dating for the extinction of the species; whereas it was hitherto thought to have disappeared some 40,000 to 20,000 years ago.

The radiocarbon surveys have further shown that a much earlier date than assumed can be given for the arrival of man in the Balearics, where it shared a coexistence with ***Myotragus balearicus*** ; the last of the ***Myotragus*** genera.

Otherwise, the Balearic cultural sequence can be demonstrated as originating from a broad geographic quarter that can best be outlined and summarised in the following:

'Arc of Immediate Influence'

The **arc of immediate influence** for Balearic prehistory is best described as a hypothetical geographic arc which can be traced, beginning in the southeast of the Iberian Peninsula in the region of Almeria and running along the Valencian coast into Catalonia and northward into the Pyrenees, Languedoc, Provence and Bouche-du-Rhone (**Figure 12A**). It is within this geographic arc that the overwhelming evidence for influence, trade and migration to the Balearics can be traced by artefact and chronometric parallels and other correlations throughout all of their cultural prehistory. A list of radiocarbon dates (**Tables 13 and 14**) both for Iberian prehistory and Balearic counterparts are presented for comparative purposes.

As pointed out in the text, the argument of comparable artefact evidence has been presented in a number of recent publications (**also see text: Parts II and III**). The chronological/chronometric evidence is argued and presented there. In the author's mind, these complement one another. In short, the criteria of the fifth and sixth stage of interpretation summarised above have been met and are manifested in the described arc of immediate influence which appeared in the Balearics as sporadic cultural waves.

The hypothetical sequence of the 'cultural waves' is as follows:

(1) **The First Cultural Wave**, circa 5000 b.c. or earlier, marks the start of our **Early Settlement Period**. As the radiocarbon survey shows in the Muleta and Matge deposits, a small number of individuals (perhaps fisherman) arrived probably by accident, carrying little equipment other than what would have been thought necessary for an off-shore fishing expedition by raft or boat, except that they seemed to have possessed some knowledge of farming and animal management. They may also have found that hunting ***Myotragus*** was an easy matter, and yielded a sufficient meat supply and clothing to support their number at first;

although not necessarily sufficient as their number grew. It is possible that the lack of ceramic technology, established by radiocarbon at circa 3000 b.c., may have been due to this arrival of the islands' first settlers having taken place much earlier than our present working date of 5000 b.c.: if so, what we see at Matge, where the earliest evidence is found, is a late phase in these first arrivals, one which still maintained an aceramic status. The other possibility being that the lower levels which produced the human remains and activity evidence at Muleta and Matge were in themselves poor.

(2) **The Second Cultural Wave**, circa 3000 b.c. This is the **NECP** of the **Pretalayotic Period**. The people were settlers who arrived having a knowledge of animal husbandry and ceramic technology, as demonstrated by radiocarbon analyses in Muleta and Matge, and who came bringing stock animals with them. This second arrival suggests higher navigational skills, and a firm intention on the part of the settlers to discover and occupy new land, whether they knew of the Balearic Islands or not.

(3) **The Third Cultural Wave**, circa 2000 b.c. Settlers and traders came to the islands bearing with them newer and more advanced pottery technology, advanced navigational skills, commerical connections, a knowledge of metal working and even architectural skills which were to bring the indigenous earlier settlers out of the caves and rock shelters; although these appear to have been used for occasional habitation by shepherd groups and other use for all of the prehistoric period. The third wave is the subject of the **EBP** of the **Pretalayotic Period**. Dwelling seems to have been largely still in caves and rock shelters at least in the intitial stages of the phase. We know now, due to extensive dating of associated archaeological contexts, that the first open-air settlements occurred at this time (**Table 14**). One of the main motives of settlement during this wave may have been metal prospection, suggested by the fact that so far most of the Bell Beaker cultural evidence and supporting radiocarbon documentation originates from the mountainous regions of Mallorca (**Figure 11**).

(4) **The Fourth Cultural Wave**, circa 1700 b.c. Commercial contacts and trade, navigation and communication with the Iberian Mainland and southern French regions seems to mark this phase until circa 1400 b.c. This **LBP** of the **Pretalayotic Period** seems to be a time when the circulation of bronze had increased considerably over the introductory phase of the EBP, and while it does appear to be a single surge of new influence as in the earlier arrivals, there appears to be a much steadier flow of contact with the mainland, especially toward 1400 b.c. with the Argaric Culture. Such minor waves of influence seem to continue

throughout the rest of the Pretalayotic Period, leading into the local Talayotic Bronze Age, although here the boundaries and understanding is less distinctive than in the preceding ones.

(5) **The Fifth Cultural Wave**, circa 1300 b.c. Based on recent datings and other findings, this fifth wave appears to have been another important in the form of the arrival of a Talayotic building culture and increased activity during this phase toward megalithic construction. There is also indications of increased population. It is here in the **EBA** of the **Talayotic Period** that a major adjustment has been made in the former frameworks of the Pentapartite Division of Balearic Prehistory in a later survival of Pretalayotic traditions and hence a later assessment for the dates of the Talayotic Culture. Toward the end of this period, the **LBA** of the **Talayotic Period**, there is an increasing influence from several directions: those of northern Europe in the form of the Hallstatt Culture as well as classical world trade of circa 800 b.c.

(6) **The Sixth Cultural Wave**, circa 800 b.c. The first iron has been introduced by the **EIA** of the **Post Talayotic Period**. Hallstatt C and D influences can be recognised in the artefacts of this phase as well later La Tene. The first classical trade items such as wheel turned pottery appears toward 400 b.c., or **MIA** of the **Post Talayotic Period**, probably due to contact with Ibiza, which was founded as a Carthaginian colony in 654 B.C. Influences and changes continue at increasing intervals until the Roman Colonisation. These influences eventually brought about the climax and seeds of decline and ultimate collapse of the Talayotic culture; although evidence shows that life in the indigenous **LIA** of the **Post Talayotic Period** settlements went on until the Visigothic arrivals.

By way of an ultimate conclusion, I believe that the current results found in this publication has shown in a positive light how successful extensive chronological and chronometric dating surveys can be in general as well as useful in delineating specific problems encountered in the interpretion of prehistoric events. While it is understood that these aspect are seen by the author as an enthusiastic and optimistic view, he also realises they can be in some respects dangerous ones in maintaining such viewpoints. For changes are bound to be forthcoming, and these are to be expected, and the whole perspective of the prehistory of this area may possibly alter still once again. Regardless, I optimistically believe that the bulk of the fund of data and information as it is set down in the present work are accurate assessments and will remain of value to other researchers for some time to come.

In all events, the actual dates as they stand will serve as reference and as a basis for recalibration when that task is undertaken, as well as forming incorporative data in the formation of any radically different Balearic prehistoric chronological scheme to come. And as stated earlier in the text, it is an accepted possibility that when that task of recalibration of the Balearic radiocarbon and other analyses results does take place, large adjustments will undoubtedly have to be made.

In the meantime, the present work is presented for the value and data it contains.

Oxford, March 1985

W.H.W.

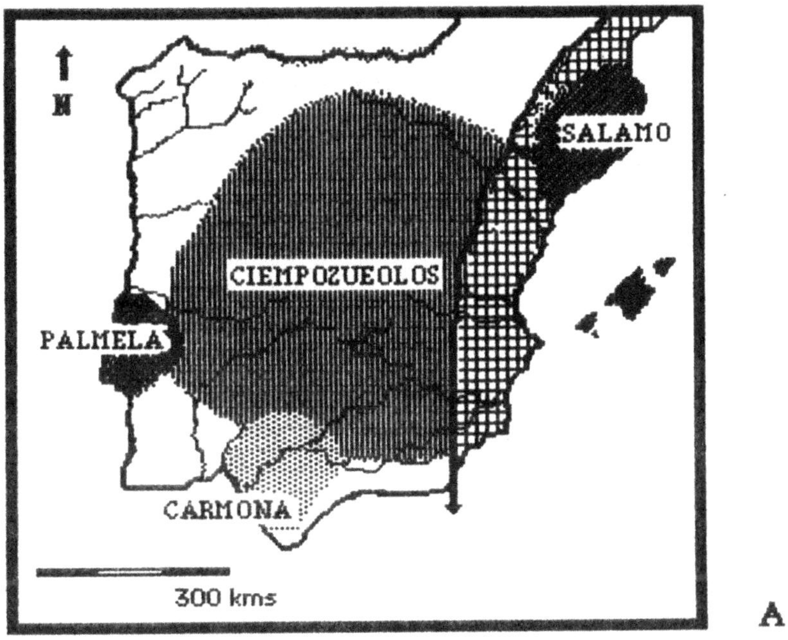

The Balearic Islands and the ARC OF IMMEDIATE INFLUENCE in Relationship to the Distribution of Iberian Bell Beaker Cultural Groups.

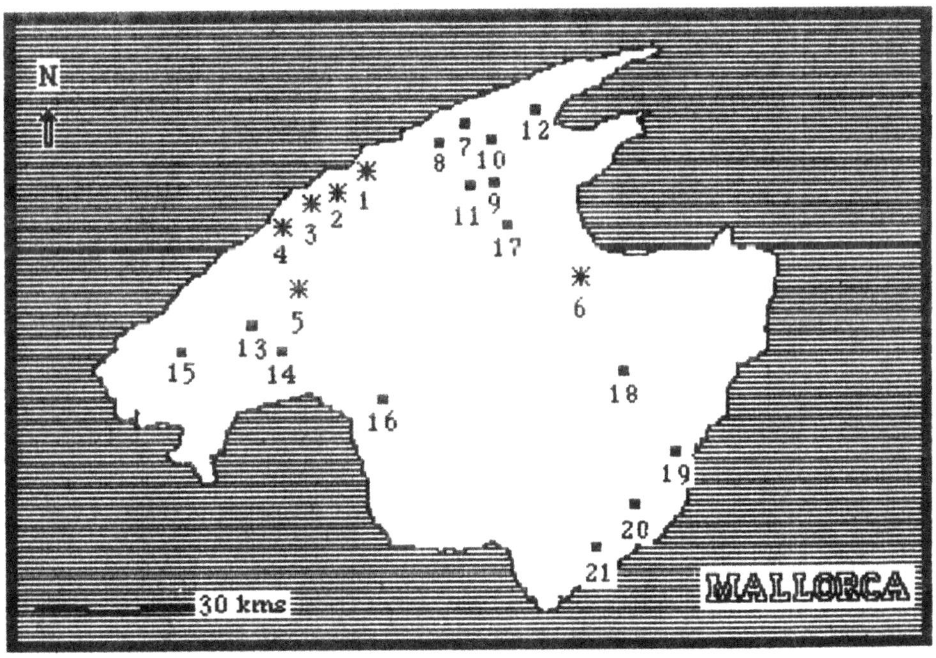

B

Distribution of Balearic Beaker Stations with and without Radiocarbon Documentation. ✳ 14c Sites

cms

FIGURE

12

CARBON 14 RESULTS
FOR THE
BALEARIC PRETALAYOTIC PERIOD
AS RELATED TO THE BELL BEAKER CULTURE

SITE/AREA	CONTEXT	bc 14C DATE	bp DATE	LAB.
		THE EARLY SETTLEMENT PERIOD		
1. ABSM	ESP	4730±120 bc	6680 yrs.	QL-29
2. SM	ESP	3985±109 bc	5935 yrs.	KBN-640c
3. ABSM	ESP	3870±360 bc	5820 yrs.	CSIC-176
4. ABSM	ESP	3800±115 bc	5750 yrs.	I-5516
		THE PRETALAYOTIC PERIOD		
		THE NEOLITHIC EARLY CERAMIC PHASE		
5. AMG	ESP	2810± 50 bc	4760 yrs.	BM-1994
6. ABSM	ESP	2700±120 bc	4650 yrs.	QL-988
7. ABSM	ESP	2143±398 bc	4093 yrs.	BM-1408
		THE EARLY BEAKER PHASE		
8. ABSM	EBP	2070± 50 bc	4020 yrs.	QL-23
9. ABSM	EBP	2030±120 bc	3980 yrs.	CSIC-178
10. ABSM	EBP	2020±100 bc	3970 yrs.	QL-5b
11. SFO-OS	EBP	2000± 65 bc	3950 yrs.	BM-1843
12. SM	EBP	1960±120 bc	3910 yrs.	Y-2389
13. ABSM	EBP	1870±120 bc	3820 yrs.	Y-2389
14. SFO-OS	EBP	1840± 90 bc	3790 yrs.	QL-1636
15. AMG	EBP	1840± 80 bc	3790 yrs.	Y-1789
16. CX	EBP	1800±120 bc	3750 yrs.	I-5515
17. SFO-OS	EBP	1770± 35 bc	3720 yrs.	BM-1981
18. SFO-OS	EBP	1750± 30 bc	3700 yrs.	QL-1592
19. ABSM	EBP	1720±100 bc	3670 yrs.	QL-24
		THE LATE BEAKER PHASE		
20. ABSM	LBP	1620± 80 bc	3620 yrs.	CSIC-179
21. SFO-OS	LBP	1540± 80 bc	3490 yrs.	QL-1859
22. ABSM	LBP	1530± 80 bc	3480 yrs.	CSIC-180
23. SMRG	LBP	1520± 80 bc	3470 yrs.	Y-1856
24. ABSM	LBP	1470± 80 bc	3420 yrs.	QL-5a
25. ABSM	LBP	1430± 50 bc	3380 yrs.	BM-1995
26. ABSM	LBP	1400± 60 bc	3350 yrs.	QL-5
27. CNC	LBP	1320± 80 bc	3270 yrs.	BM-1667
28. SFO-OS	LBP-EBA	1260± 80 bc	3210 yrs.	BM-2312
29. ABSM	LBP-EBA	1250±100 bc	3200 yrs.	Y-2667
30. SFO-OS	LBP-EBA	1140± 70 bc	3090 yrs.	BM-1698
31. SFO-OS	LBP-EBA	1120± 50 bc	3070 yrs.	HAR-3490

TABLE 13

A SELECTION OF IBERIAN BEAKER RADIOCARBON DATES

SITE	14C DATE	BP DATE
1. Zambugal IVb	2075 ± 95 b.c.	4025 yrs.
2. Los Millares	2070 ± 60 b.c.	4020 yrs.
3. Zambugal IIIb	2040 ± 35 b.c.	3990 yrs.
4. Zambugal IVb	2000 ± 65 b.c.	3950 yrs.
5. Ereta del Pedregal	1980 ± 250 b.c.	3930 yrs.
6. Los Huesos	1970 ± 100 b.c.	3920 yrs.
7. Virgen de Orce	1970 ± 60 b.c.	3920 yrs.
8. Montefrio V	1890 ± 35 b.c.	3840 yrs.
9. Virgen de Orce	1887 ± 35 b.c.	3847 yrs.
10. Virgen de Orce	1850 ± 35 b.c.	3800 yrs.
11. Terlinques	1850 ± 115 b.c.	3800 yrs.
12. Los Millares	1820 ± 60 b.c.	3870 yrs.
13. Praia des Macas	1700 ± 100 b.c.	3650 yrs.
14. Cabezo Redondo	1600 ± 55 b.c.	3550 yrs.
15. Pic dels Corbs	1581 ± 100 b.c.	3531 yrs.
16. Cati Forada	1552 ± 50 b.c.	3393 yrs.
17. Mas de Abad	1460 ± 90 b.c.	3410 yrs.
18. Cabezo Redondo	1350 ± 55 b.c.	3300 yrs.

A SELECTION OF SOUTHERN FRENCH RADIOCARBON DATES

SITE	14C DATE	BP DATE
17. Lebous		
18. Bois Sacre	1940 + 140 b.c.	3930 yrs.
19. Camp Laure	1710 + 100 b.c.	3660 yrs.
20. Camp Laure	1550 + 60 b.c.	3500 yrs.

TABLE

14

BALEARIC BELL BEAKER RADIOCARBON CHRONOLOGY

MALLORCA

MATGE

rock shelter stratigraphy

2700 ± 120 yrs.
(pre-beaker)

2070 ± 50 yrs.
2030 ± 120 yrs.
2020 ± 100 yrs.
1870 ± 120 yrs.
1720 ± 100 yrs.
1620 ± 80 yrs.
1540 ± 80 yrs.
1470 ± 80 yrs.
1430 ± 50 yrs.
1400 ± 60 yrs.
1250 ± 100 yrs.

FERRANDELL-OLEZA

open-air stratigraphy

2000 ± 65 yrs.
1840 ± 90 yrs.
1770 ± 35 yrs.
1750 ± 30 yrs.
1540 ± 80 yrs.
1260 ± 80 yrs.
1140 ± 70 yrs.
1120 ± 50 yrs.

MUERTOS-GALLARD

rock shelter stratigraphy

2810 ± 50 yrs.
(pre-beaker)

1840 ± 80 yrs.

COTXERA

open-air stratigraphy

1800 ± 115 yrs.

MARROIG

cave stratigraphy

1520 ± 80 yrs.

FORMENTERA

CA NA COSTA

open-air stratigraphy

1320 ± 80 yrs.

* all 14C dates uncalibrated

TABLE 15

BIBLIOGRAPHY

<<<<<<<<<<<<>>>>>>>>>>>

Adrover, Hno., R. 1966. 'Pequeno Intento de Lavado de la Cueva de Son Muleta y los Resultados Obtenido'. **Boletin de la Sociedad de Historia Natural de Baleares, XII**, Palma de Mallorca, pp. 38-47.

Adrover, Hno., R. 1977. 'Baleafrensis Multetensis, a New Species of Frog'. **Boletin de la Sociedad de Historia Natural de Baleares, XII**, Palma de Mallorca, pp.49-51

Alcover y Tomas, J. 1976. 'L'Evolucio de Myotragus Bate 1909. Un Proces Biologiic Lligat al Fenomen de la Insularitat' **Barcelona Instituto de Ciencia, Historia Natural, 40**, pp. 59-94.

Almagro Gorbea, M. 1978. 'C14 y Prehistoria de la Peninsula Iberica'. **Seria Universitaria**, Fundacion March, Madrid.

Altschuler, Z.S., Clarke, R.S. and Young, E.J., 1958. '**Geochemistry of Uranium in Apatite and Phosphonite**', U.S. Geological Survey Professional Paper, 314-D.

Amoros, Amoros, L. 1955. 'Sa Canova, Petra, Mallorca'. **Noticiario Arquelogico Hispano 2**, Madrid, pp.17.

Andrews, C.W., 1914. 'The Description of a Skull and Skeleton of a Particularly Modified Rupicaprine Antelope' (Myotragus balearicus with a Notice of a New Variety), **Philosophical Transactions of the Royal Society of London**, Vol. 206, Series B, pp.281-305.

Atkinson, T.C., Harmon, R.S., Smart, P.L. and Waltham, A.C., 1978. 'Palaeoclimatic and Geomorphic Implications of 230Th/234U Dates on Speleothems from Britain', **Nature 272**, pp. 24-28.

Arribas, Palau A. 1963. 'El Poblado y la Necropolis Megalithicos de los Millares'. **Biblioteca Prehistorica Hispana, vol III**, Madrid.

Bada, J.L. and Schroeder, R.A. 1973. 'Glacial-Postglacial Temperature Differences Deduced from Aspartic Acid Racemisation in Fossil Bones', **Science 2**, November, Vol. 182, pp. 479-482.

Baillie,M.G.L. and Pilcher, J.R. 1983. 'Archaeology Dendrochronology and the Radiocarbon Calibration Curve', editor, B.S. Ottaway, **University of Edinburgh**, Occasional Paper No.9, Edinburgh, pp. 51-63.

Bate, D.M. 1909. 'A New Artiodactyla from Mallorca', **Geological Magazine**, New Series, Decade V, Vol. VI. London, pp. 385-388.

Bate, D.M. 1918. 'On the Pleistocene Ossiferous Deposits of the Balearic Islands', **Geological Magazine**, Decade VI, Vol. I, London, pp.337-345.

Bate, D.M. 1919. 'On a New Genera of Extinct Muscardine Rodent from the Balearic Islands'. **Geological Magazine**, New Series, Decade VI, Vol. II, London, Pp. 342-352.

Boekschoten, G.L. and Sondaar, P.Y. 1972. ' On the Fossil Mammalia of Cyprus'. **Knok, Nederlands Akademie van Wetenschappen, Proceedings**. Series B, Vol.75, No. 4, Amsterdam, pp 306-338.

Burleigh, R. and Clutton-Brock, J. 19 80. ' The Survival of Myotragus balearicus Bate, 1909 in the Neolithic of Mallorca'. **Journal of Archaeological Sciences 7**, pp. 285-288.

Camps Cantarellas, C. 1972. 'Cerámica Incisa en Mallorca'. **Caja de Ahorros y Monte Piedad de las Baleares**, Palma de Mallorca.

Castillo, Yurrita, A. 1928. "La Cultura de Vaso Campaniforme'. **Universidad de Barcelona, Barcelona**.

Clark, R.M.,1975. 'A Calibration Curve for Radiocarbon Dates', **Antiquity XLIX**, pp. 251-256.

Colom, Roca, G. 1957. 'Bibliografia de las Baleares.' **Estudio General Luliana**. Palma de Mallorca.

Colom, Roca, G. 1975. ' Sobre los Nuevos Nociones Generales Sobre la Evolucion Paleografica y Poblamiento del Archipielago Balear desde el Eocene al Cuaternario'. **Revista Balear 28-39**, Palma de Mallorca, pp. 7-34.

Colominas, Roca, J. 1915. 'Sobre L'Edad de Bronce a Mallorca'. **Anuari del'Institut d'Estudis Catalans 5**, Barcelona.

Crusafont, Pairo, M. and Angel, Basilio, Hno., 1966. 'Un Myotragus, Mamifere Ruminat, dans le Villafranchien de l'ile de Mallorca: Myotragus batei, nov. sp.'. **Centre de Reserche. Academie de Science. Paris**, pp. 2014-2015.

Cuerda, J. 1966 and 1974. ' Los Tiempos Cuaternarios en las Baleares'. **Sociedad de Historia Natural de Baleares**, Palma de Mallorca.

Dehaut, E.G. 1911 and 1920. 'Les Antilopes Rupicaprines Pleistocene de l'iles Baleares et de las Saraignia'. **Contribucio Pulbicada a les Contributions a l'Etude de la Vie Vertebree Insulaire dans la Region Mediterraneene Occidentale et Particulierment en Sardaigne et Corse**, Paris.

Damon, P.E., Long, A. and Wallick E.I., 1972. Dendrochronological Calibrationof the Radiocarbon Time Scale, **Proceedings of the 8th Int'l Conference on Radiocarbon Dating**, Lower Hutt City, I, pp.45-59.

Duplessy, J.C., Labeyrie, J., Lalou, C. and Nguyen, H.V., 1970. **Nature 226**, pp.

Ensenat Estrany, B., 1963. Noticias Sobre el Hallazgo en Mallorca de unas Ceramicas Incisas de Estilo de Vaso Campaniforme'. **Congreso Nacional de Arquelogia.** Barcelona. pp. 184.

Evans, j.d., 1973. ' Islands as Laboratories of Cultural Change', in C. Renfrew (ed.), **Explanation of Cultural Change, Models in Prehistory**, Duckworth, London, pp. 517-520.

Fernandez Gomez, J., Plantalamor Massanet, L. and Topp, C. 1974. 'Excavaciones en el Sepulcor Megalitico de Ca Na Costa. Formentera'. **Estudios de Prehistoria Balear 1**, Palma de Mallorca.

Fernandez, J., Plantalamor Massanet, L. and Topp, C., 1976. Excavaciones en el Sepulcor Megalitico de Ca Na Costa, Separata de Mayudga, 15. **Estudios de Prehistoria Balear**. Ministerior de Educación y Ciencia, Palma de Mallorca, pp. 109-138.

Fernandez Gomez, J. and Topp, C. 1984. 'Prehistoric Activity in the Pitiussae Islands'. The Deya Conference of Prehistory ..Early Settlement in the Western Mediterranean Islands and their Peripheral Areas. **BAR International Series, Oxford. vol. iii**, pp.763-783.

Fernandez-Miranda, M. 1979. 'Secuencia Cultural de la Prehistoria de Mallorca'. **Biblioteca Prehistorica Hispana**, Madrid.

Fernandez-Miranda, M. and Waldren, W.H., 1974. 'El Abrigo de Son Matge, (Valldemosa) y la Periodización de la Prehistoria Mallorquina mediante las Análasis de Carbon 14'. **Trabajos de Prehistoria 31**. Madrid, pp. 287-304.

Fernandez-Miranda, M. and Waldren, W.H., 1979. 'Periodificación Cultural y Cronológia Absoluta en la Prehistoria de Mallorca'. **Trabajos de Prehistoria 36**. Madrid, pp. 349-377.

Gasull, P., Lull, V. and Sanahuja, C. 1983.

Gilman, A., 1981 'The Development of Social Stratification in Bronze Age Europe', **Current Anthropology 22**, pp. 169-224.

Gottesfeld, M., Martin, P. and Waldren, W.H. 1968. 'Fossil Pollen Counts from the Cave of Muleta, Mallorca'. **DAMARC Series 5**, Deya de Mallorca, Baleares, Spain.

Guilaine, J. 1967. 'La Civilation du Vaso Campaniforme dans le Pyrennees Francaise'. **Centre National de Researche Scientifique**, Carcassone.

Guilaine, J. 1970. 'L' Age du Bronze en Languedoc Occidentale, Pousillon, Ariege'. **Klincksieck**, Paris.

Gutherz, X. 1978. 'Informations Archeologiques' **Gallia Prehistoire 21**, pp. 678-690.

Gutherz, X. and Hugues, C.., 1980, 'La Culture du Vase Campaniforme dans le Departement du Gard Ecre Antique de Nimes, **Nimes**, pp. 5-26.

Harrison, R.J. 1980. 'The Beaker Folk, Copper Age Archaeology in Western Europe'. **Thames and Hudson**, London.

Harrison, R.J. 1974. 'Origin of the Bell Beaker Cultures'. **Antiquity**, London, pp. 99-109.

Hemp, W. 1932. 'The Naveta of Menorca', **Antiquaries Journal 12**, London, pp. 127-135.

Ivanovich, M. 1982. Spectroscopic Methods, in M. Ivanovich and R.S. Harmon (eds.) **Uranium Series Disequilbrium Applications to Environmental Problems in the Earth Sciences**, Oxford University Press, Oxford, pp. 107-144.

Jacobsen, T.W. 1976. 'Excavations in the Franchthi Cave, 1969-1971'. **Hesperia 42**, pp. 45-88 and 253-283.

Lilliu, G. 1972. 'La Civilta dei Sardi da Neolitico all Eta dei Nuraghi'. **Annali dell Facolta de Lettree dell Universita di Cagliari**, Sardinia.

Lilliu, G. 1962. 'Cenno sui piu Recenti Scavi de Villago Talayotico de Ses Paisses, Mallorca'. **Annali dell Facolta de Lettree dell Universita di Cagliari 27**, Sardinia, pp. 1-44.

Little, K., 1973. **Bone Behaviour**, Academic Press, London.

Mascaro, Pasarius, J. 1967. 'Corpus de Toponima de Mallorca'. **Graficas Miramar V**, Palma de Mallorca.

Mapper, D., Bolus, D.J. and Stephen, J., 1982. The Application of Uranium Fission Track Autoradiography as a Method of Investigating the Problem of Soft Errors in VLSI Silicon Memory Devices, in P.H. Fowler and V.M. Clapham (eds.) **Solid State Nucleur Track Detectors**, Pergamon, Oxford, pp.815-823.

Mills, D. 1974. 'Hypnomys morpheus: an Osteological Study and Comparative Rodentia'. **Swedish National Museum**, Stockholm.

Morales, A. 1979. 'C14 y Prehistoria de la Peninsula Iberica'. **Seria Universitaria**, Madrid, pp. 65-6.

Murray, M. 1938. ' The Cambridge Expeditions'. **B. Quaritch 11**, London.

Neuman, W.F., Neuman, M.W., Main, E.P. and Mulryan, B.J., 1949. The Deposition of Uranium Bone, **J. Biol. Chem. 179**, pp. 325-348.

Pericot, Garcia, L. 1973. ' The Balearic Islands'. **Thames and Hudson**, London.

Plantalmor Massanet, L. and Waldren, W.H. 1976. 'Campana de Excavaciones en el Abrigo de Son Matge. **N. A. H. Prehistoria 5**, Madrid, pp. 243-252.

Plantalamor Massanet, L. 1981 'Algunas Consideraciones Sobre los Sepulcros Megaliticos de Menorca'. Separata de Sautuola 11, **Trabajos del Museo de Menorca**. Mahon de Menorca.

Pons, Moya, J. 1977. 'La Nouvelle Espece Myotragus antiquss de l'Ile de Majorque, Baleares. **Koninklijke Nederlands Akademie van Vetenschappen**, Amsterdam, Series B. Vol. 80, pp.215-221.

Ralph, E.K. and Michael, H.N., 1970. MASCA Radiocarbon Dates for Sequoia and Bristlecone Pine Samples, **Radiocarbon Variations and Absolute Chronology** (I.U. Olason, Ed.), pp. 100-138.

Renfrew, C., 1973. Before Civilisation: the Radiocarbon Revolution and Prehistoric Europe, **Jonathan Cape**, London.

Renfrew, C. and Clark, R.M., 1974. Problems of the Radiocarbon Calendar and Its Calibration, **Archaeometry 16**, pp. 5-18.

Reumer J.W.F. ,1980. 'On Two New Micromamals from the Pleistocene of Mallorca'. **Geological Institute, University of Utrecht**, Utrecht, pp. 473-482.

Romer, A.S. 1962. 'Vertebrate Palaeontology'. **University of Chicago Press**, Chicago, pp.473-482.

Rossello Bordoy, G. 1958. 'Cerámica Incisa en Mallorca'. **Studi Sardi, IV and XVI**, Cagliari.

Rossello Bordoy, G. 1960-1961. 'La Facies 'Cueva Natural' en la Cultura Pretalayotica Mallorquina'. **Ampurias**, pp. 263.

Rossello Bordoy, G. 1962. 'Excavaciones en la Necropolis de Cuevas Artificiales de Son Sunyer'. **Excavaciones Arqueologicas en Espana 14**, Madrid.

Rossello Bordoy, G. and Lilliu, G., 1963. Cenno sui piu Recenti Scavi nel Villaggio Talayotico di Ses Paisses ad Mallorca en SS 18, **Cagliari**, pp. 22-48.

Rossello Bordoy, G. and Frey, H., 1964. Ein Talayot-Seidlung se S'Illot, Mallorca, **Madrider Mitteilungen 5**, Madrid.

Rossello Bordoy, G. 1964. 'Ses Salinas, Avance al Estudio de sus Yacimientos Arqueologicas'. **Boletin de la Sociedad Arqueologica Luliana XXIII**, Palma de Mallorca.

Rossello Bordoy, G., Kopper, J.S. and Waldren, W.H. 1967. Complejo Norde de Son Figueral de Son Real, Santa Margarita, Mallorca. **I Congreso Internacionale de Arqueologia**, Mahon y Saragosa, pp. 83-85.

Rossello Bordoy, G., 1973. La Cultura Talayotica en Mallorca: Bases para el Estudio de sus Fases Iniciales. **Caja de Ahorros y Monte de Piedad de las Baleares**, Palma de Mallorca.

Rossello Bordoy, G. and Waldren, W.H. 1973. 'Excavaciones en el Abrigo de Bosque de Son Matge, Valldemosa'. **N.A.H. Prehistoria 11**, p. 1-86.

Rossello Bordoy, L., Plantalamor Massanet, L. and Pons Lopez, A. 1980. 'Excavaciones Arqueológicos en Torre den Gaumes, Alayor, Menorca. La Sepultura Megalitica de Ses Roques Llises', **N. A. H. Prehistoria 8**, Madrid, pp. 71-138.

Schule, W., 1981. 'El Cerro de la Virgen, Orce (Granada), **Excavaciones Arqueologicas en España 46**, Madrid

Schwarz, H.P. 1980. Uranium Series Dating of Archaeological Sites, **Archaeometry 22**, pp. 3-38.

Schwarz, H.P. 1982. Application of U-Series Dating to Archaeometry, in M. Ivanovich and R.S. Harmon (eds.) **Uranium Series Disequilbrium Applications to Environmental Problems in the Earth Sciences**, Oxford University Press, pp. 326-350.

Schwarz, H.P. , Gascoyne, M. and Harmon, R.S. 1982. Applications of U-Series Dating to Problems of Quaternary Climate, in M. Ivanovich and R.s. Harmon (eds.), **Uranium Desies Disequilbrium Applications to Environmental Problems in the earth Sciences**, Oxford University Press, pp.302-325.

Seitz, M.G. and Taylor, R.E., 1974. Uranium Variations in a Dated Fossil Bone Series from Olduvai Gorge, Tanzania, **Archaeometry 16**, pp. 129-135.

Sondaar, p.y. 1977. 'Insularity and Its Effect on Mammal |Evolution', in M.K. Hecht, P.C. Boody and B.M. Hecht (eds.), **Major Patterns in Vertebrate Evolution**, Plenum Publishing Corporation, New York, pp. 671-707

Stuiver, M. and Waldren, W.H., 1974. 14C Carbonate Dating and the Age of Post Talayotic Lime Burials in Mallorca. Nature, Vol. 255, June, pp. 475-476.

Suess, H.E., 1970. Bristlecone Pine Calibration of the Radiocarbon Time Scale from 5400 b.c. to the Present, **Radiocarbon Variations and Absolute Chronology** (I.U. Olason, Ed.), pp. 303-312.

Switzur, J., 1973. in Clark's "A Calibration Curve for Radiocarbon Dates", **Antiquity, XLIX**, pg. 251.

Szabo, B.J. and Collins, D., 1975. Ages of Fossil Bone from British Interglacial Sites, **Nature** 254, pp. 679-681.

Szabo, B.J., Malde, H.E. and Irwin-Williams, C., 1969. Dilemma Posed by Uranium Series Dates on Archaeologically Significant Bones from Valsequillo, Puebla, Mexico, **Earth Plant. Sci. Letters 6**, pp. 237-244.

Tarradell, M. and Woods, D., 1976. Excavaciones en la Necropolis de Son Real y Ille dels Porros. **Excavaciones Arqueologicas en Espana 24**, Madrid.

Veny, C. 1961. 'Cueva del Tossal Verts:, Malbice, **Sartabi XI**, Valencia, pp. 201.

Veny, C. 1968. 'Las Cuevas Sepulcrales del Bronce Antiguo de Mallorca. Consejo Superior de Investigaciones Cientificas'. **Instituto de Prehistoria Espanol de Madrid**, Madrid.

Waldren, W.H. 1967. 'Informe Preliminar Sobre Analisis de Radiocarbono en Mallorca'. **X Congreso Nacional de Arqueologia. Mahon y Zaragossa**, pp. 75-82.

Waldren, W.H. 1968. 'Mallorcan Chronology for Prehistory Based on Radiocarbon Method'. **Pyrenae 3**, Barcelona, pp. 45-65.

Waldren, W.H. 1969. 'Beaker Ware from the Island of Mallorca'. **DAMARC Series No. 6**, Deya de Mallorca, Baleares.

Waldren, W.H. 1972. 'Age Determination by Radiocarbon Method'. **Sociedad de Historia Natural de las Baleares**, Palma de Mallorca, pp. 34-50.

Waldren, W.H. 1973. 'Excavations of the Rock Shelter of Son Matge, Valldemosa'. **Prehistoria II**, Madrid.

Waldren, W.H. 1974. 'Prehistory and the Archaeology of the Balearic Islands'. **VI Symposium of Peninsular Prehistory**, University of Barcelona, pp. 31-38.

Waldren, W.H. 1979. 'A Beaker Workshop Area in the Rock Shelter of Son Matge, Valldemosa, Mallorca' **World Archaeology**, Routledge and Kegan Paul, London, Vol 11, No. 1, pp. 43-67.

Waldren, W.H. 1980. Radiocarbon Determination in the Balearic Islands: an Inventory, **Donald Baden-Powell Quaternary Research Centre**, Oxford University, Oxford.

Waldren, W.H. 1980. 'Radiocarbon Determination in the Balearic Islands; An Inventory', **Donald Baden Powell Quaternary Research Centre, Oxford University**, Oxford.

Waldren, W.H. 1980. 'The Settlement Complex of Ferrandell-Oleza, A Beaker Settlement from the Balearic Island of Mallorca', **Donald Baden Powell Quaternary Research Centre, Oxford University, Oxford and DAMARC Series 8**, Deya de Mallorca, Baleares, Spain.

Waldren, W.H. 1981. 'Preliminary Report on the Settlement Complex of Ferrandell-Oleza, Valldemosa, Mallorca. (in preparation) **DAMARC Series 10**, Deya de Mallorca, Baleares.

Waldren, W.H. 1982. Aspects of Balearic Prehistoric Ecology and Culture. **Oxford Doctoral Thesis**, Oxford.

Waldren, W.H. 1982. 'Balearic Prehistoric Ecology and Culture: The Excavation and Study of Certain Caves, Rock Shelters and Settlements', **BAR International Series 149** (ii) Appendix 3A, pp. 675-711. Oxford.

Waldren, W.H. 1983. 'Early Prehistoric Settlement in the Balearic Islands'. **DAMARC Series 13**, Deya Archaeological Museum and Research Centre, Deya Mallorca.

Waldren, W.H. 1984. ' Chalcolithic Settlement and Beaker Connections in the Balearic Islands'. The Deya Conference of Prehistory ...Early Settlement in the Western Mediterranean Islands and their Peripheral Areas. **BAR International Series 229** , Oxford. vol. iii, pp.911-965.

Waldren, W.H. 1984. ' The Lithic Industry of the Balearic Islands....its Olezian Tradition of Tabular Flint Blades'. The Deya Conference of Prehistory ...Early Settlement in the Western Mediterranean Islands and their Peripheral Areas. **BAR International Series 229** , Oxford. vol. iii, pp.859-909.

V

THE AGE DETERMINATIONS AND OTHER INVENTORIES

This section is dedicated to the **Inventory of Radiocarbon Analyses Results** listings and tables related to the individual analytical dating methods used (**e.g Racemisation of Aspartic Acid Analysis of Bone and Uranium Thorium Dating of Bone**), also included are the **Analyses Results Index Cards** for more detailed reference to each of the listed analyses.

A brief summary of the recording system used in the inventory has been outlined in the Introduction. Although the inventory listing is self explanatory, some repetition of that summary is necessary here for easier and immediate reference.

In the Inventory of Radiocarbon Analyses Results listing tables, the first heading is the inventory number (**e.g. 5, 5a, 5b, 5c, etc.**). These as pointed out earlier are arranged so that an in-coming analysis result can be slotted into the body of the inventory. This is done for convenience sake in chronological/chronometrical order of age. The second heading is the particular result's Balearic Island designation of origin (**e.g Me: Minorca, Ma: Mallorca, etc.**). The third heading is the Site Origin designation (**e.g. ABSM: Rock Shelter of Son Matge, SFO-OS: Ferrandell Oleza Old Settlement, etc.**). The fourth heading is the Official Facility Laboratory Number (**e.g. BM: British Museum, QL: Quaternary Research Centre, University of Washington at Seattle, etc.**). The fifth heading is the Test Material Used (**Ch: Charcoal, Ab: Animal Bone, etc.**). The sixth heading is Period Horizon or Level (**e.g. ESP: Early Settlement Period, PT: Post Talayotic, etc.**). The last heading is the Test Sample Age in Years Before the Present (**b.p.**) and Age Before and After Christ (**b.c., a.d.**).

References throughout the author's text are designated in the following manner: Site abbreviation (**e.g. ABSM: Rock Shelter of Son Matge**) then the Inventory number (**47a, 47b, etc.**). When combined the radiocarbon reference number will appear in the following manner (**ABSM 47a etc.**)

All dates are as quoted throughout the text are uncalibrated. These uncalibrated radiocarbon dates are given in lower case alphabetical reference (**a.d.** or **b.c.**). Calibrated and historical dates are given in capital alphabetical reference (**A.D.** or **B.C.**). A **5568 yrs half-life** is recognised for the radiocarbon method.

Analyses Results Index Cards are included here and give specifics concerning the collection, the materials, the sample weights, year of the analysis and the scientist conducting the individual tests and other pertinent observations.

A list of **Abbreviations for University, Institution and Laboratory Facilities, Test Materials, Sites and Sample Origins** are provided at the end of the Inventory.

The Balearic Quantitative Spectrographic Bronze Analysis Series

The Balearic Islands have very few if any <u>extensive</u> copper resources and no known tin. And although some poor copper sources have been reported, it is doubtful whether or not the copper that does exist is sufficient in quantity to have fostered a prosperous prehistoric copper working industry. Yet the prehistoric archaeological contexts from the Pretalayotic Period through the Talayotic and Post Talayotic Periods have produced rich bronze items, such as swords, daggers, knives, large pectoral decorations, bracelets and rings and other ornaments. Admittedly most of the really elaborate artefacts appear chronologically quite late, circa 1000 b.c. forward; however copper items and copper working are know as early as 2000 b.c. and there is some evidence of an early presence of bronze, circa 1700 b.c. Actual evidence of metal working does exist in the form of copper interior encrusted Beaker, decorated and undecorated crucible sherds and modest caches of copper awls and a few small ingot clippings (**Waldren 1979**).

In order to establish the sources of the Balearic raw materials as well as its compositions, a Balearic quantitative, spectrographic bronze analysis series was initiated in 1978. For this purpose the first series of 6 samples from Mallorca sites were analysed by Dr. Peter J. Northover, Department of Metallurgy, Oxford University, with the results shown below, at the end of the radiocarbon inventory

A second more extensive series of 45 samples were further analysed in 1985, with the results in table form at the end of the radiocarbon inventory.

Both these series form part of a larger Iberian survey currently underway. No attempt, however, at present is made to synthesise in any way these results. This process will have to be undertaken at a future date. It will suffice to say the results are bioth interesting and important to the area and that a further extension of the series is presently planned. The more detailed publication dedicated solely to this subject of Balearic metal objects and their sources is in the planning stage. For the moment, the analyses results todate are presented here so as to further the function of this present publication, as one offering the analyses results available in the Balearic Islands here in 1986.

Inventory
Radiocarbon Analyses Results
Including
Racemisation of Aspartic Acids Analyses Results
in
Chronological Order of Age

No.	L.	Site	Lab. No.	M.	Level	Age
			Post Colonisation and Post Talayotic Periods			
1.	Ma	AMG	**BM-1993**	Ch.	M	885 yrs. b.p. **1145 ± 35 a.d.**
1a.	Me	TT	**BM-2005**	Ch.	PC	1560 yrs. b.p. **350 ± 80 a.d.**
1b.	Ma	SFO-OS	**QL-1860**	Ch.	PC	1650 yrs. b.p. **300 ± 15 a.d.**
1c.	Ma	SFO-OS	**QL-1858**	Ch.	PC	1670 yrs. b.p. **280 ± 15 a.d.**
1d.	Ma	SFO-OS	**BM-1982**	Ch.	PC	1710 yrs. b.p. **240 ± 60 a.d.**
1e.	Ma	SFO-OS	**QL-1861**	Ch.	PC	1730 yrs. b.p. **220 ± 12 a.d.**
1f.	Me	TT	**BM-2004**	Ch.	PC	1890 yrs. b.p. **60 ± 35 a.d.**
1g.	Me	SRDC	**QL-146**	Ca.	PT	2030 yrs. b.p. **80 ± 40 b.c.**
2.	Ma	ABSM	**QL-7a**	Ca.	PT	2070 yrs. b.p. **120 ± 120 b.c.**
3.	Ma	ABSM	**QL-8**	Ch.	PT	2080 yrs. b.p. **130 ± 90 b.c.**
3a.	Me	TT	**BM-2003**	Ch.	PT	2090 yrs. b.p. **140 ± 50 b.c.**
3b.	Me	TT	**CSIC-?**	Ch.	PT	2100 yrs. b.p. **150 ± 45 b.c.**
3c.	Ma	SFO-OS	**BM-2297**	Ch.	PT	2140 yrs. b.p. **160 ± 80 b.c.**
4.	Ma	SM	**SI-651a**	Ch.	PT	2180 yrs. b.p. **230 ± 115 b.c.**
5.	Me	TT	**CSIC-142**	Ab.	PT	2180 yrs. b.p. **230 ± 100 b.c.**
6.	Ma	SPG	**Y-2673**	Hb.	PT	2180 yrs. b.p. **230 ± 100 b.c.**
7.	Ma	ABSM	**QL-9**	Ca.	PT	2200 yrs. b.p. **250 ± 100 b.c.**
8.	Ma	AMG	**Y-2672**	Ch.	PT	2230 yrs. b.p. **280 ± 100 b.c.**

TABLE 16

9.	Me	ABSM	**QL-1a**	Ca.	PT	2240 yrs. b.p. **290 ± 70 b.c.**
10.	Me	ABSM	**QL-22**	Ca.	PT	2260 yrs. b.p. **310 ± 60 b.c.**
11.	Me	SPA	**CSIC-?**	Ch.	PT	2270 yrs. b.p. **320 ± 110 b.c**
12.	Me	ABSM	**QL-5c**	Ch.	PT	2290 yrs. b.p. **340 ± 100 b.c**
12a.	Me	SO	**BM-1692**	Cg.	PT	2290 yrs. b.p. **340 ± 40 b.c.**
12b.	Me	SB	**BM-1517**	Wd.	PT	2350 yrs. b.p. **400 ± 35 b.c.**
13.	Me	SMAR	**QL-144**	Ch.	PT	2370 yrs. b.p. **420 ± 50 b.c.**
13a.	Me	SB	**BM-1518**	Wd.	PT	2390 yrs. b.p. **440 ± 45 b.c.**
13b.	Me	SFO-YS	**I-5398**	Ch.	PT	2400 yrs. b.p. **450 ± 60 b.c.**
14.	Me	ABSM	**Y-2669**	Ca.	PT	2400 yrs. b.p. **450 ± 80 b.c.**
15.	Me	SRN	**I-4584**	Ch.	PT	2430 yrs. b.p. **480 ± 200 b.c.**
15a.	Me	SFO-YS	**BM-1842**	Ab.	PT	2430 yrs. b.p. **480 ± 230 b.c.**
16.	Me	SRDC	**QL-145**	Ch.	PT	2450 yrs. b.p. **500 ± 40 b.c.**
16a.	Me	FORN	**UGRA-121**	Ch.	PT	2450 yrs. b.p. **500 ± 190 b.c.**
16b.	Me	SFO-YS	**HAR-3459**	Ch.	PT	2400 yrs. b.p. **510 ± 80 b.c.**
16c.	Me	FORN	**UGRA-123**	Ch.	PT	2470 yrs. b.p. **520 ± 80 b.c.**
17.	Me	ABSM	**QL-10**	Ch.	PT	2480 yrs. b.p. **530 ± 70 b.c.**
18.	Me	SO	**Y-2666**	Ch.	PT	2490 yrs. b.p. **540 ± 40 b.c.**
18a.	Me	FORN	**UGRA-122**	Ch.	PT	2490 yrs. b.p. **540 ± 130 b.c.**
18b.	Fo	CNC	**BM-1510**	Ab.	PT	2500 yrs. b.p. **550 ± 100 b.c.**
19.	Me	ABSM	**QL-6**	Ch.	PT	2520 yrs. b.p. **570 ± 80 b.c.**
19a.	Me	FORN	**UGRA-382**	Ch.	PT	2540 yrs. b.p. **590 ± 85 b.c.**
19b.	Me	FORN	**UGRA-124**	Ch.	PT	2540 yrs. b.p. **590 ± 80 b.c.**
19c.	Me	SFO-YS	**HAR-3458**	Ch.	PT	2540 yrs. b.p. **590 ± 60 b.c.**
20.	Me	ABSM	**QL-24**	Ch.	PT	2540 yrs. b.p. **590 ± 80 b.c.**
21.	Me	ABSM	**QL-4**	Ch.	PT	2540 yrs. b.p. **590 ± 100 b.c.**
21a.	Me	FORN	**UGRA-381**	Ch.	PT	2560 yrs. b.p. **610 ± 85 b.c.**

TABLE

16a

22.	Ma	ABSM	**QL-20**	Ch.	PT	2570 yrs. b.p. **620 ± 100 b.c.**
22a.	Ma	TBC	**HAR-2891**	Ch.	PT	2570 yrs. b.p. **620 ± 100 b.c.**
23.	Ma	ABSM	**QL-27**	Ch.	PT	2640 yrs. b.p. **690 ± 100 b.c.**
23a.	Ma	SSP	**BM-1998**	Ch.	PT	2645 yrs. b.p. **695 ± 40 b.c.**
23b.	Fo	CNC	**BM-1511**	Hb.	PT	2670 yrs. b.p. **720 ± 60 b.c.**
24.	Ma	SILL	**HU-1718**	Ch.	PT	2690 yrs. b.p. **740 ± 60 b.c.**
24a.	Ma	FORN	**UGRA-123**	Ch.	PT	2700 yrs. b.p. **750 ± 170 b.c.**
25.	Ma	ABSM	**QL-11**	Ch.	PT	2700 yrs. b.p. **750 ± 80 b.c.**
25a.	Ma	TT	**QL-1165**	Ch.	PT	2710 yrs. b.p. **760 ± 50 b.c.**
26.	Ma	ABSM	**QL-7**	Ch.	PT	2730 yrs. b.p. **780 ± 100 b.c.**
			Talayotic Period			
27.	Ma	SM	**SI-552**	Gb.	T	2765 yrs. b.p. **815 ± 120 b.c.**
27a.	Ma	NSM	**HAR-2909**	Ch.	T	2770 yrs. b.p. **820 ± 100 b.c.**
28.	Ma	ABSM	**QL-986**	Ch.	T	2820 yrs. b.p. **860 ± 50 b.c.**
29.	Me	TT	**QL-1164**	Ch.	T	2830 yrs. b.p. **880 ± 40 b.c.**
29a.	Me	TT	**QL-1089**	Ch.	T	2840 yr. b.p. **890 ± 30 b.c.**
29b.	Me	TT	**BM-1697**	Ch.	T	2860 yrs. b.p. **910 ± 45 b.c.**
30.	Ma	SP	**GIF-1247**	Ch.	T	2900 yrs. b.p. **950 ± 120 b.c.**
30a.	Ma	SFO-YS	**HAR-3413**	Ch.	T	2910 yrs. b.p. **960 ± 100 b.c.**
31.	Ma	SO	**QL-20**	Ch.	T	2920 yrs. b.p. **970 ± 60 b.c.**
32.	Ma	SRF	**Y-1857**	Ch.	T	2960 yrs. b.p. **1010 ± 120 b.c.**
33.	Ma	SILL	**HU-1717**	Ch.	T	2960 yrs. b.p. **1010 ± 110 b.c.**
33a.	Me	TT	**HAR-2908**	Ch.	T	2970 yrs. b.p. **1020 ± 70 b.c.**
34.	Ma	SRF	**Y-1857**	Ch.	T	3000 yrs. b.p. **1050 ± 120 b.c.**
34a.	Me	TT	**HAR-?**	Ch.	T	3020 yrs. b.p. **1070 ± 60 b.c.**
34b.	ME	TT	**QL-1433**	Cg.	T	3030 yrs. b.p. **1080 ± 70 b.c.**

TABLE

16b

34c	Ma	SFO-OS	**HAR-3490**	Ab	T-PRT	3070 yrs. b.p. **1120 ± 50 b.c.**
35.	Ma	SILL	**HU-1716**	Ch.	T	3080 yrs. b.p. **1130 ± 75 b.c.**
35a.	Ma	SFO-OS	**BM-1698**	Ab.	T-PRT	3095 yrs. b.p. **1145 ± 70 b.c.**
36	Ma	ABSM	**Y-2667**	Ch.	T	3200 yrs. b.p. **1250 ± 100 b.c.**
36a.	Ma	SFO-OS	**BM-2312**	Ab	T-PRT	3210 yrs. b.p. **1260 ± 80 b.c.**
37	Ma	SP	**UP-1438**	Ch.	T	3260 yrs. b.p. **1310 ± 60 b.c.**
37a.	FO	CNC	**BM-1677**	Hb.	PRT	3270 yrs. b.p. **1320 ± 80 b.c.**
			Pretalayotic Period			
38.	Ma	ABSM	**QL-5**	Cahl.	PRT	3350 yrs. b.p. **1400 ± 60 b.c.**
38a.	Ma	ABSM	**BM-1995**	Chcl.	PRT	3380 yrs. b.p. **1430 ± 50 b.c.**
39	Ma	ABSM	**QL-5a**	Cahl.	PRT	3420 yrs. b.p. **1470 ± 100 b.c.**
40.	Ma	SMRG	**Y-1856**	Hb.	PRT	3470 yrs. b.p. **1520 ± 80 b.c.**
41.	Ma	ABSM	**CSIC-180**	Chcl.	PRT	3480 yrs. b.p. **1530 ± 80 b.c.**
41a.	Ma	SFO-OS	**QL-1859**	Ab.	PRT	3490 yrs. b.p. **1540 ± 80 b.c.**
42.	Ma	ABSM	**CSIC-179**	Chcl.	PRT	3620 yrs. b.p. **1620 ± 80 b.c.**
43.	Ma	ABSM	**QL-24**	Cahl.	PRT	3670 yrs. b.p. **1720 ± 100 b.c.**
43a.	Ma	SFO-OS	**QL-1592**	Chcl.	PRT	3700 yrs. b.p. **1750 ± 30 b.c.**
43b.	Ma	SFO-OS	**BM-1981**	Chcl.	PRT	3270 yrs. b.p. **1770 ± 35 b.c.**
44.	Ma	CX	**I-5515**	Ch.	PRT	3750 yrs. b.p. **1800 ± 120 b.c.**
45.	Ma	AMG	**Y-1789**	Ch.	PRT	3790 yrs. b.p. **1840 ± 80 b.c.**
45a.	Ma	SFO-OS	**QL-1636**	Ch.	PRT	3790 yrs. b.p. **1840 ± 90 b.c.**
46.	Ma	ABSM	**Y-2359**	Ch	PRT	3820 yrs. b.p. **1870 ± 120 b.c.**
47.	Ma	SM	**Y-2359**	Ch.	PRT	3910 yrs. b.p. **1960 ± 120 b.c.**
47a.	Ma	SFO-OS	**BM-1843**	Ch.	PRT	3950 yrs. b.p. **2000 ± 60 bc.**
48.	Ma	ABSM	**QL-5b**	Cahl.	PRT	3970 yrs. b.p. **2020 ± 100 b.c.**
49.	Ma	ABSM	**CSIC-178**	Chhl.	PRT	3980 yrs. b.p. **2030 ± 170 b.c.**

TABLE

16c

50.	Ma	ABSM	**QL-23**	Chhl.	PRT	4020 yrs. b.p. **2070 ± 50 b.c.**
50a.	Ma	ABSM	**BM-1408**	Mb.	PRT	4093 yrs. b.p. **2143 ± 392 b.c.**
51.	Ma	ABSM	**QL-988**	Chhl.	PRT	4650 yrs. b.p. **2700 ± 120 b.c.**
51a.	Ma	AMG	**BM-1994**	Ch.	PRT	4760 yrs. b.p. **2810 ± 50 b.c.**
			Early Settlement Period			
52.	Ma	ABSM	**I-5516**	Chhl.	ESP	5750 yrs. b.p. **3800 ± 115 b.c.**
53.	Ma	ABSM	**CSIC-176**	Cps.	ESP	5820 yrs. b.p. **3870 ± 360 b.c.**
54.	Ma	SM	**KBN-640d**	Hb.	ESP	5935 yrs. b.p. **3985 ± 109 b.c.**
55.	Ma	ABSM	**QL-29**	Mbhl.	ESP	6680 yrs. b.p. **4730 ± 120 b.c.**
			Presettlement Period			
56.	Ma	SM	**KBN-640c**	Mb.	PSP 0150cms	7135 yrs. b.p. **5185 ± 80 b.c.**
56a.	Ma	SM	**UCLA-1704j**	Mb.	PSP	7300 yrs. b.p. **6350 ± 100 b.c.**
56b.	Ma	SM	**HAR-2934**	Mb.	PSP X7c 65cms	16,500 yrs. b.p. **15,500 ± 2300 b.c.**
57.	Ma	SM	**UCLA-1704c**	Mb.	PSP 0150cms	8570 yrs. b.p. **6020 ± 350 b.c.**
57a.	Ma	SM	**UCLA-1704f**	Mb.	PSP X100cms	8800 yrs. b.p. **6850 ± 400 b.c.**
57b.	Ma	SM	**HAR-2909**	Mb.	PSP	16,400 yrs. b.p. **14,450 ± 1200 b.c.**
57c.	Ma	SM	**HAR-2932**	Mb.	PSP X7E 100 cms	19,700 yrs. b.p. **17,750 ± 1110 b.c.**
58.	Ma	SM	**KBN-640b**	Mb.	PSP 0150cms	8448 yrs. b.p. **6498 ± 1180 b.c.**
59.	Ma	SM	**KBN-640c**	Mb.	PSP 0175cms	10,685 yrs. b.p. **8735 ± 3517 b.c.**
60.	Ma	SM	**SIO-1f**	Mb.	PSP Z175cms	(pending)
61.	Ma	SM	**SI-654a**	Mb.	PSP Z350cms	14,000 yrs. b.p. **12,050 ± 350 b.c.**
62.	Ma	SM	**SI-654b**	Mb.	PSP Z350cms	14,465 yrs. b.p. **12,515 ± 315 b.c.**
63.	Ma	SM	**UCLA-170b**	Mb.	PSP Z250cms	14,650 yrs. b.p. **12,700 ± 850 b.c.**
64.	Ma	SM	**SI-646**	Mb.	PSP D250cms	15,885 yrs. b.p. **13,985 ± 655 b.c.**
64a.	Ma	SM	**HAR-2933**		PSP X11	23,500 yrs. b.p. **21,550 ± 1600 b.c.**
65.	Ma	SM	**SI-648**	Mb.	PSP	16,335 yrs. b.p.

TABLE

16d

					X300cms	**14,385 ± 415 b.c.**
66.	Ma	SM	**UCLA-1704d**	Mb.	PSP	16,850 yrs. b.p.
					E350cms	**14,900 ± 200 b.c.**
66a.	Ma	SM	**HAR-2931**		PSP	14,200 yrs. b.p.
					X13	**12,250 ± 1100 b.c.**
67.	Ma	SM	**SI-649**	Mb.	PSP	18,100 yrs. b.p.
					F300cms	**16,150 ± 600 b.c.**
68.	Ma	SM	**SIO-1i**	Mb.	PSP	
					Z400cms	(pending)
69.	Ma	SM	**SI-650**	Mb.	PSP	18,735 yrs. b.p.
					X350cms	**16,805 ± 555 b.c.**
70.	Ma	SM	**UCLA-1704e**	Mb.	PSP	18,980 yrs. b.p.
					Z400cms	**17,030 ± 200 b.c.**
71.	Ma	SM	**SI-647**	Mb.	PSP	23,800 yrs. b.p.
					E400cms	**21,850 ± 480 b.c.**
71a.	Ma	SM	**HAR-2912**	Mb.	PSP	23,200 yrs. bp.
					E355 cms	**21,250 ± 2300 b.c.**
72.	Ma	SM	**SIO-1c**	Mb.	PSP	26,000 yrs. b.p.
					Z475cms	**25,050 ± 600 b.c.**
72a.	Ma	SM	**HAR-2910**		PSP	25,800 yrs. b.p.
					Z555cms	**23,850 ± 1300 b.c.**
73.	Ma	SM	**UCLA-1704a**	Mb.	PSP	28,600 yrs. b.p.
					Z600cms	**26,650 ± 600 b.c.**
74.	Ma	SM	**SIO-1g**	Mb.	PSP	32,000 yrs. b.p.
					Z750cms	**30,050 ± 715 b.c.**
74a.	Ma	SM	**SIO-1h**	Mb.	PSP	42,000 yrs. b.p.
					Z600cms	**40,050 ± 1000 b.c.**
74b.	Ma	SM	**HAR-2911**	Mb.	PSP	44,800 yrs. b.p.
					Z650cms	**42,850 ± 1600 b.c.**
75.	Ma	SM	**QL-127**	Sc.	PSP	48,000 yrs. b.p. (exterior)
					X base	48,000 yrs. (infinite)
76.	Ma	SM	**QL-128**	Sc.	PSP	58,000 yrs. b.p. (interior)
					X base	58,000 yrs. (infinite)
77.	Ma	SM	**Y-1f**	Cf.	PSP	80,000 yrs. b.p. RF
					X base	80,000 yrs.

TABLE

16e

LIST OF ABBREVIATIONS

Universities, Institutions and Laboratories

BM	British Museum, Radiocarbon Laboratory, London.
CSIC	University of Madrid, Centro Superior de Investigaciones Cientifica, Spain.
GIF	French Radiocarbon Laboratory.
HU	German Radiocarbon Laboratory.
I	Isotopes Incorporated, Englewood, New Jersey.
KBN	Kaman Bio-Nuclear Inc., Colorado Springs, Colorado.
QL	Quaternary Research Centre, University of Washington Radiocarbon Laboratory, Seattle Washington.
SI	Smithsonian Institution, United States National Museum, Radiocarbon Laboratory, Washington, D.C.
SIO	Scripps Institute of Oceanography, University of California, La Jolla, California.
UCLA	University of California at Los Angeles, Radiocarbon Laboratory, Los Angeles, California.
UGRA	University of Granada, Granada, Spain.
UP	University of Pennsylvania, Museum School, Radiocarbon Laboratory, Philadelphia, Pennsylvania.
Y	Yale University, Peabody Museum, Radiocarbon Laboratory, New Haven, Connecticut.

Test Material Abbreviations

Ab.	Animal bone
Ca.	Carbonate
Cahl.	Carbonate from hearth levels
Cf.	Calcite floor
Cg.	Carbonised grain
Ch.	Charcoal
Ch.b.	Charred bone
Chhl.	Charcoal from hearth levels
Cps.	Coprolites
Gb.	Goat bone
Hb.	Human bone
Mb.	Myotragus bone
Mbhl.	Myotragus bone in hearth levels
Sc.	Stalactite core
Wd.	Wood

TABLE

17

Abbreviations of Sites and Sample Origins

Ma.	Mallorca
Me.	Menorca
Fo.	Formentera
b.p.	Before the present
b.c.	Uncalibrated 14C dates
Z300	Sector and Level in centimetres at Son Muleta
PSP	Presettlement Period
ESP	Early Settlement Period
PRT	Pretalayotic
T	Talayotic
PT	Post Talayotic
ABSM	Rock Shelter of Son Matge
AMG	Rock Shelter of Muertos Gallard
BCA	Talayot of Binicalf
CX	Ca Na Coxtera
FORN	Son Fornes
SB	Son Baronat
SFO-YS	Ferrandell-Oleza Older Settlement
SFO-OS	Ferrandell-Oleza Younger Settlement
SILL	S'illot
SM	Cave of Son Muleta
SMAR	Son Marino
SMR	Naviforme of Son Morell
SMRG	Cave of Son Marroig
SO	Son Oms
SP	Son Severa-Pula
SPG	Cave of Son Puig
SRDC	Sa Regina dels Cans
SRF	Son Real Figureral
SRN	Son Real Necropolis
SSP	Ses Paisses
TT	Taula of Torralba den Salort

TABLE

18

DAMARC — Dating Analyses Index Cards

Material: Charcoal

Coll. Date: 1984

Lab. No.: BM-1993

Observations:

Laboratory Facility: British Museum

Uncalibrated Dates: 885 yrs. b.p. 1145 ± 35 a.d.

BRCP Calibrated Dates:

Site: Rock Shelter of Muertos Gallard, Mallorca

Site Abbreviation: AMG

Inventory Number: 1

DAMARC — Dating Analyses Index Cards

Material: Charcoal

Coll. Date: 1984

Lab. No.: BM-2005

Observations:

Laboratory Facility: British Museum

Uncalibrated Dates: 1560 yrs. b.p. 350 ± 15 a.d.

BRCP Calibrated Dates:

Site: Taula de Torralba den Salort, Menorca

Site Abbreviation: TT

Inventory Number: 1a

DAMARC — Dating Analyses Index Cards

Material:	Charcoal
Coll. Date:	1984
Lab. No.:	QL-1860
Laboratory Facility:	University of Washington
Uncalibrated Dates:	1650 yrs. b.p. 300 ± 15 a.d.
BRCP Calibrated Dates:	

Observations

Charcoal from upper levels, probably the result of field fires of recent times and field clearance activities

Site:	Ferrandell-Oleza Old Settlement, Mallorca
Site Abbreviation:	SFO-OS
Inventory Number:	1b

DAMARC — Dating Analyses Index Cards

Material:	Charcoal
Coll. Date:	1982
Lab. No.:	QL-1858
Laboratory Facility:	University of Washington
Uncalibrated Dates:	1670 yrs b.p. 280 ± 15 a.d.
BRCP Calibrated Dates:	

Observations

Charcoal probably like sample 1b from field fires and clearance activities.

Site:	Ferrandell-Oleza Old Settlement, Mallorca
Site Abbreviation:	SFO-OS
Inventory Number:	1c

DAMARC — Dating Analyses Index Cards

Material: Charcoal

Coll. Date: 1983

Lab. No.: BM-1982

Observations

Charcoal from field clearance fires from upper level at the base of a modern terrace wall.

Laboratory Facility: British Museum

Uncalibrated Dates: 1710 yrs. b.p. 240 ± 60 a.d.

BRCP Calibrated Dates:

Site: Ferrandell-Oleza Old Settlement, Mallorca

Site Abbreviation: SFO-OS

Inventory Number: 1d

DAMARC — Dating Analyses Index Cards

Material: Charcoal

Coll. Date: 1983

Lab. No.: QL-1861

Observations

Charcoal at the base of a modern terrace wall, probably from a clearance fire.

Laboratory Facility: University of Washington

Uncalibrated Dates: 1730 yrs. b.p. 220 ± 12 a.d.

BRCP Calibrated Dates:

Site: Ferrandell-Oleza Old Settlement, Mallorca

Site Abbreviation: SFO-OS

Inventory Number: 1e

DAMARC | Dating Analyses Index Cards

Material: Charcoal

Coll. Date: 1984

Lab. No.: BM-2004

Observations

Charcoal from upper level inside Taula probably a very late occupation level.

Laboratory Facility: British Museum

Uncalibrated Dates: 1890 yrs. b.p. 60 ± 35 a.d.

BRCP Calibrated Dates:

Site: Taula de Torralba den Salort, Menorca

Site Abbreviation: TT

Inventory Number: 1f

DAMARC | Dating Analyses Index Cards

Material: Quicklime

Coll. Date: 1981

Lab. No.: QL-146

Observations

Quicklime burial sample from a cave on Minorca. C. Veny excavation.

Laboratory Facility: University of Washington

Uncalibrated Dates: 2030 yrs. b.p. 80 ± 40 b.c.

BRCP Calibrated Dates:

Site: Sa Regina dels Cans, Menorca

Site Abbreviation: SRDC

Inventory Number: 1g

DAMARC — Dating Analyses Index Cards

Material:	Quicklime
Coll. Date:	1972
Lab. No.:	QL-7a
Laboratory Facility:	University of Washington
Uncalibrated Dates:	2070 yrs. b.p. 120 ± 120 b.c.
BRCP Calibrated Dates:	

Observations

Part of original quicklime burial dating series

Site:	Rock Shelter of Son Matge, Mallorca
Site Abbreviation:	ABSM
Inventory Number:	2

DAMARC — Dating Analyses Index Cards

Material:	Charcoal
Coll. Date:	1982
Lab. No.:	QL-8
Laboratory Facility:	University of Washington
Uncalibrated Dates:	2070 yrs. b.p. 130 ± 90 b.c.
BRCP Calibrated Dates:	

Observations

Charcoal related to quicklime burials.
Part of original dating series.

Site:	Rock Shelter of Son Matge, Mallorca
Site Abbreviation:	ABSM
Inventory Number:	3

DAMARC — **Dating Analyses Index Cards**

Material:	Charcoal
Coll. Date:	1983
Lab. No.:	BM-2003
Laboratory Facility:	British Museum
Uncalibrated Dates:	2090 yrs. b.p. 140 ± 50 b.c.
BRCP Calibrated Dates:	

Observations

Charcoal from occupational level in the Taula structure associated with classical artefacts.

Site:	Taula de Torralba den Salort, Menorca
Site Abbreviation:	TT
Inventory Number:	3a

DAMARC — **Dating Analyses Index Cards**

Material:	Charcoal
Coll. Date:	1981
Lab. No.:	CSIC-?
Laboratory Facility:	University of Madrid
Uncalibrated Dates:	2100 yrs. b.p. 150 ± 45 b.c.
BRCP Calibrated Dates:	

Observations

Charcoal from occupation level similar to Sample 3 from inside the Taula.

Site:	Taula de Torralba den Salort, Menorca
Site Abbreviation:	TT
Inventory Number:	3b

DAMARC — Dating Analyses Index Cards

Material:	Charcoal
Coll. Date:	1984
Lab. No.:	BM-2297
Laboratory Facility:	British Museum
Uncalibrated Dates:	2140 yrs. b.p. 160 ± 80 b.c.
BRCP Calibrated Dates:	

Observations

Charcoal from upper level. Probably from field clearance activities after the Old Settlement abandonment.

Site: Ferrandell-Oleza Old Settlement, Mallorca

Site Abbreviation: SFO-OS

Inventory Number: 3c

DAMARC — Dating Analyses Index Cards

Material:	Charcoal
Coll. Date:	1966
Lab. No.:	SI-651a
Laboratory Facility:	Smithsonian Institution
Uncalibrated Dates:	2180 yrs. b.p. 230 ± 100 b.c.
BRCP Calibrated Dates:	

Observations

Charcoal from small kitchen hearths in the Cave of Muleta. Fires associated with domesticated animals, goat, sheep etc.

Site: Cave of Son Muleta, Mallorca

Site Abbreviation: SM

Inventory Number: 4

DAMARC — Dating Analyses Index Cards

Material: An. Bone

Coll. Date: 1981

Lab. No.: CSIC-142

Observations

Domesticated animal bone samples from beneath benches in northeast sector.

Laboratory Facility: University of Madrid

Uncalibrated Dates: 2180 yrs. b.p. 230 ± 100 b.c.

BRCP Calibrated Dates:

Site: Taula de Torralba den Salort, Menorca

Site Abbreviation: TT

Inventory Number: 5

DAMARC — Dating Analyses Index Cards

Material: H. Bone

Coll. Date: 1966

Lab. No.: Y-2673

Observations

Human bone from the burial cave of Son Puig. Associated with some classical pottery. However sample was originally thought to be of Pretalayotic origin.

Laboratory Facility: Yale University

Uncalibrated Dates: 2180 yrs. b.p. 230 100 b.c.

BRCP Calibrated Dates:

Site: Cave of Son Puig, Mallorca

Site Abbreviation: SPG

Inventory Number: 6

DAMARC	Dating Analyses Index Cards
Material:	Quicklime
Coll. Date:	1969
Lab. No.:	QL-9
Observations	Part of original quicklime burial survey dating at Matge.
Laboratory Facility:	University of Washington
Uncalibrated Dates:	2200 yrs. b.p. 250 ± 100 b.c.
BRCP Calibrated Dates:	
Site:	Rock Shelter of Son Matge, Mallorca
Site Abbreviation:	ABSM
Inventory Number:	7

DAMARC	Dating Analyses Index Cards
Material:	Charcoal
Coll. Date:	1968
Lab. No.:	Y-2672
Observations	First lime samples to be dated along with the first at Matge.
Laboratory Facility:	Yale University
Uncalibrated Dates:	2230 yrs. b.p. 280 ± 100 b.c.
BRCP Calibrated Dates:	
Site:	Rock Shelter of Muertos Gallard, Mallorca
Site Abbreviation:	AMG
Inventory Number:	8

DAMARC — Dating Analyses Index Cards

Material:	Quicklime
Coll. Date:	1969
Lab. No.:	QL-1a
Observations	Part of original quicklime burial survey at Matge.
Laboratory Facility:	University of Washington
Uncalibrated Dates:	2240 yrs. b.p. 290 ± 70 b.c.
BRCP Calibrated Dates:	
Site:	Rock Shelter of Son Matge, Mallorca
Site Abbreviation:	ABSM
Inventory Number:	9

DAMARC — Dating Analyses Index Cards

Material:	Quicklime
Coll. Date:	1969
Lab. No.:	QL-22
Observations	Part of original Quicklime burial dating survey at Matge
Laboratory Facility:	University of Washington
Uncalibrated Dates:	2260 yrs. b.p. 310 ± 60 b.c.
BRCP Calibrated Dates:	
Site:	Rock Shelter of Son Matge
Site Abbreviation:	ABSM
Inventory Number:	10

DAMARC

Dating Analyses Index Cards

Material: Charcoal

Coll. Date: ?

Lab. No.: CSIC

Observations

Laboratory Facility: University of Madrid

Uncalibrated Dates : 2270 yrs. b.p. 320 ± 110 b.c.

BRCP Calibrated Dates:

Site: SES PAISSES, MALLORCA

Site Abbreviation: SPA

Inventory Number: 11

DAMARC

Dating Analyses Index Cards

Material: Charcoal

Coll. Date: 1969

Lab. No.: QL-5c

Observations

Part of original quichlime burial survey at Matge.

Laboratory Facility: University of Washington

Uncalibrated Dates : 2290 yrs. b.p. 340 ± 100 yrs.

BRCP Calibrated Dates:

Site: Rock Shelter of Son Matge, Mallorca

Site Abbreviation: ABSM

Inventory Number: 12

DAMARC — Dating Analyses Index Cards

Material:	Car. Grain
Coll. Date:	1980
Lab. No.:	BM-1692
Laboratory Facility:	British Museum
Uncalibrated Dates:	2290 yrs. b.p. 340 ± 40 b.c.
BRCP Calibrated Dates:	

Observations

Carbonised grain from the Talayotic settlement of Son Oms - Posttalayotic origin.

Site:	Son Oms, Mallorca
Site Abbreviation:	SO
Inventory Number:	12a

DAMARC — Dating Analyses Index Cards

Material:	Wood
Coll. Date:	1980
Lab. No.:	BM-1517
Laboratory Facility:	British Museum
Uncalibrated Dates:	2350 yrs. b.p. 400 ± 35 b.c.
BRCP Calibrated Dates:	

Observations

Site:	Son Baronat, Mallorca
Site Abbreviation:	SB
Inventory Number:	12b

DAMARC — Dating Analyses Index Cards

Material: Charcoal

Coll. Date: 1978

Lab. No.: QL-144

Observations

Laboratory Facility: University of Washington

Uncalibrated Dates: 2370 yrs. b.p. 420 ± b.c.

BRCP Calibrated Dates:

Site: Son Marino, Mallorca

Site Abbreviation: SMAR

Inventory Number: 13

DAMARC — Dating Analyses Index Cards

Material: Wood

Coll. Date: 1980

Lab. No.: BM-1518

Observations

Wood from a Post Talayotic sarcophagus Sample is similar to 12b above. Wood identified as larch by Kew Gardens.

Laboratory Facility: British Museum

Uncalibrated Dates: 2390 yrs. b.p. 440 ± 45 b.c.

BRCP Calibrated Dates:

Site: Son Baronat, Mallorca

Site Abbreviation: SB

Inventory Number: 13a

DAMARC — Dating Analyses Index Cards

Material: Charcoal

Coll. Date: 1981

Lab. No.: I-5398

Observations

Charcoal from the abandonment level of Talayot *2 of the Younger Settlement. Samples in Oxford for species identification.

Laboratory Facility: Isotopes Inc, USA

Uncalibrated Dates: 2400 yrs. b.p. 450 ± 60 b.c.

BRCP Calibrated Dates:

Site: Ferrandell-Oleza Younger Settlement, Ma.

Site Abbreviation: SFO-YS

Inventory Number: 13b

DAMARC — Dating Analyses Index Cards

Material: Quicklime

Coll. Date: 1969

Lab. No.: Y-2669

Observations

Part of Quicklime burial survey from Son Matge.

Laboratory Facility: Yale University

Uncalibrated Dates: 2400 yrs. b.p. 450 ± 80 b.c.

BRCP Calibrated Dates:

Site: Rock Shelter of Son Matge, Mallorca

Site Abbreviation: ABSM

Inventory Number: 14

DAMARC — Dating Analyses Index Cards

Material: Charcoal

Coll. Date: 1966

Lab. No.: I-4584

Observations

Laboratory Facility: Isotopes Inc, USA

Uncalibrated Dates: 2430 yrs. b.p. 480 ± 200 b.c.

BRCP Calibrated Dates:

Site: Necropolis of Son Real, Mallorca

Site Abbreviation: SRN

Inventory Number: 15

DAMARC — Dating Analyses Index Cards

Material: An. Bone

Coll. Date: 1983

Lab. No.: BM-1842

Observations

Animal bone from kitchen fires in T°1
Talayot °1 (part of series).

Laboratory Facility: British Museum

Uncalibrated Dates: 2430 yrs. b.p. 480 ± 230 b.c.

BRCP Calibrated Dates:

Site: Ferrandell-Oleza Younger Settlement, Ma

Site Abbreviation: SFO-YS

Inventory Number: 15a

DAMARC — Dating Analyses Index Cards

Material:	Charcoal
Coll. Date:	1978
Lab. No.:	QL-145
Observations:	
Laboratory Facility:	University of Washington
Uncalibrated Dates:	2450 yrs. b.p. 500 ± 40 b.c.
BRCP Calibrated Dates:	
Site:	Son Regina dels Cans, Mallorca
Site Abbreviation:	SRDC
Inventory Number:	16

DAMARC — Dating Analyses Index Cards

Material:	Charcoal
Coll. Date:	1980?
Lab. No.:	UGRA-121
Observations:	Part of V. Lull Son Fornes series.
Laboratory Facility:	University of Granada
Uncalibrated Dates:	2450 yrs. b.p. 500 ± 190 b.c.
BRCP Calibrated Dates:	
Site:	Son Fornes Settlement, Mallorca
Site Abbreviation:	FORN
Inventory Number:	16a

DAMARC | Dating Analyses Index Cards

Material:	Charcoal
Coll. Date:	1983
Lab. No.:	HAR-3459
Observations	
Laboratory Facility:	Harwell, England
Uncalibrated Dates :	2460 yrs. b.p. 510 ± 80 b.c.
BRCP Calibrated Dates:	
Site:	Ferrandell-Oleza Younger Settlement, Ma
Site Abbreviation:	SFO-YS
Inventory Number:	16b

DAMARC | Dating Analyses Index Cards

Material:	Charcoal
Coll. Date:	1980
Lab. No.:	UGRA
Observations	
Laboratory Facility:	University of Granada
Uncalibrated Dates :	2470 yrs. b.p. 520 ± 80 yrs.
BRCP Calibrated Dates:	
Site:	Son Fornes Settlement, Mallorca
Site Abbreviation:	FORN
Inventory Number:	16c

DAMARC — Dating Analyses Index Cards

Material:	Charcoal
Coll. Date:	1970
Lab. No.:	Q1-10
Laboratory Facility:	University of Washington
Uncalibrated Dates:	2480 yrs. b.p. 530 ± 70 b.c.
BRCP Calibrated Dates:	

Observations: Part of Quicklime burial dating series at Matge

Site: Rock Shelter of Son Matge, Mallorca

Site Abbreviation: ABSM

Inventory Number: 17

DAMARC — Dating Analyses Index Cards

Material:	Charcoal
Coll. Date:	1979
Lab. No.:	Y-2666
Laboratory Facility:	Yale University
Uncalibrated Dates:	2490 yrs. b.p. 540 ± 40 b.c.
BRCP Calibrated Dates:	

Observations:

Site: Son Oms Settlement, Mallorca

Site Abbreviation: SO

Inventory Number: 18

DAMARC — Dating Analyses Index Cards

Material: Charcoal

Coll. Date: 1980

Lab. No.: UGRA-122

Observations: Part of V. Lull's Formes series and same series as samples 16c and 16a.

Laboratory Facility: University of Granada

Uncalibrated Dates: 2490 yrs. b.p. 540 ± 130 b.c.

BRCP Calibrated Dates:

Site: Son Fornes Settlement, Mallorca

Site Abbreviation: FORN

Inventory Number: 18a

DAMARC — Dating Analyses Index Cards

Material: An. Bone

Coll. Date: 1981

Lab. No.: BM-1510

Observations: Part C. Topp, Fernandez Formentera series at Can Na Costa.

Laboratory Facility: British Museum

Uncalibrated Dates: 2500 yrs. b.p. 550 ± 100 b.c.

BRCP Calibrated Dates:

Site: Meglithic Tomb of Can Na Costa, Form.

Site Abbreviation: CNC

Inventory Number: 18b

DAMARC — Dating Analyses Index Cards

Material:	Charcoal
Coll. Date:	1979
Lab. No.:	QL-1533
Laboratory Facility:	University of Washington
Uncalibrated Dates:	2500 yrs. b.p. 550 ± 40 b.c.
BRCP Calibrated Dates:	

Observations

Date from occupation level in Talayot T1 of the Younger Settlement.

Site: Ferrandell-Oleza Younger Settlement, Ma

Site Abbreviation: SFO-YS

Inventory Number: 18c

DAMARC — Dating Analyses Index Cards

PENDING

Material:	
Coll. Date:	
Lab. No.:	
Laboratory Facility:	DESCRIPTION PENDING
Uncalibrated Dates:	
BRCP Calibrated Dates:	

Observations

Site:

Site Abbreviation:

Inventory Number: 18c

DAMARC — Dating Analyses Index Cards

Material: Charcoal

Coll. Date: 1970

Lab. No.: QL-6

Observations

Laboratory Facility: University of Washington

Uncalibrated Dates: 2520 yrs b.p. 570 ± 80 b.c.

BRCP Calibrated Dates:

Site: Rock Shelter of Son Matge, Mallorca

Site Abbreviation: ABSM

Inventory Number: 19

DAMARC — Dating Analyses Index Cards

Material: Charcoal

Coll. Date: 1980

Lab. No.: UGRA-382

Observations

Laboratory Facility: University of Washington

Uncalibrated Dates: 2540 yrs. b.p. 590 ± 80 b.c.

BRCP Calibrated Dates:

Site: Son Fornes Settlement, Mallorca

Site Abbreviation: FORN

Inventory Number: 19a

DAMARC

Dating Analyses Index Cards

Material: Charcoal

Coll. Date: 1980

Lab. No.: UGRA-124

Observations

Part of V. Lull's Fornes series.

Laboratory Facility: University of Granada

Uncalibrated Dates : 2540 yrs. b.p. 590 ± 80 b.c.

BRCP Calibrated Dates:

Site: Son Fornes Settlement, Mallorca

Site Abbreviation: FORN

Inventory Number: 19b

DAMARC

Dating Analyses Index Cards

Material: Charcoal

Coll. Date: 1983

Lab. No.: HAR-3458

Observations

Part of series dating the construction of Talayot *! (T1) in the Younger Settlement. Occupational level.

Laboratory Facility: Harwell, England

Uncalibrated Dates : 2540 yrs. b.p. 590 ± 60 b.c.

BRCP Calibrated Dates:

Site: Ferrandell-Oleza Younger Settlement, Ma

Site Abbreviation: SFO-YS

Inventory Number: 19c

DAMARC — Dating Analyses Index Cards

Material:	Charcoal
Coll. Date:	1970
Lab. No.:	QL-24
Observations	
Laboratory Facility:	University of Washington
Uncalibrated Dates:	2540 yrs. b.p. 590 ± 80 b.c.
BRCP Calibrated Dates:	

Site:	Rock Shelter of Son Matge, ‹allorca
Site Abbreviation:	ABSM
Inventory Number:	20

DAMARC — Dating Analyses Index Cards

Material:	Charcoal
Coll. Date:	1970
Lab. No.:	QL-4
Observations	
Laboratory Facility:	University of Washington
Uncalibrated Dates:	2540 yrs. b.p. 590 ± 100 b.c.
BRCP Calibrated Dates:	

Site:	Rock Shelter of Son Matge, Mallorca
Site Abbreviation:	ABSM
Inventory Number:	21

DAMARC | Dating Analyses Index Cards

Material: Charcoal

Coll. Date: 1980

Lab. No.: UGRA-381

Observations

Part of V. Lull's dating series at Son Fornes. See Samples 19a, 19b, 18a 16c.

Laboratory Facility: University of Granada

Uncalibrated Dates: 2560 yrs. b.p. 610 ± 85 b.c.

BRCP Calibrated Dates:

Site: Son Fornes Settlement, Mallorca

Site Abbreviation: FORN

Inventory Number: 21a

DAMARC | Dating Analyses Index Cards

Material: Charcoal

Coll. Date: 1970

Lab. No.: QL-20

Observations

Laboratory Facility: University of Washington

Uncalibrated Dates: 2570 yrs. b.p. 620 ± 100 b.c.

BRCP Calibrated Dates:

Site: Rock Shelter of Son Matge, Mallorca

Site Abbreviation: ABSM

Inventory Number: 22

DAMARC

Dating Analyses Index Cards

Material:	Charcoal
Coll. Date:	1981
Lab. No.:	HAR-2891
Laboratory Facility:	Harwell, England
Uncalibrated Dates:	2570 yrs. b.p. 620 ± 100 b.c.
BRCP Calibrated Dates:	

Observations

Analysis for Piantalamor on Menorca

Site:	Talayot of Binicallafa, Menorca
Site Abbreviation:	TBC
Inventory Number:	22a

DAMARC

Dating Analyses Index Cards

Material:	Charcoal
Coll. Date:	1970
Lab. No.:	QL- 27
Laboratory Facility:	University of Washington
Uncalibrated Dates:	2640 yrs. b.p. 690 ± 100 b.c.
BRCP Calibrated Dates:	

Observations

Oldest ate so far for dating quicklime burail at Matge

Site:	Rock Shelter of Son Matge, Mallorca
Site Abbreviation:	ABSM
Inventory Number:	23

DAMARC — Dating Analyses Index Cards

Material:	Charcoal
Coll. Date:	1981
Lab. No.:	BM-1998
Laboratory Facility:	British Museum
Uncalibrated Dates:	2645 yrs. b.p. 695 ± 40 b.c.
BRCP Calibrated Dates:	

Observations

Sample done for Rossello Bordoy of the Museum of Mallorca

Site: Son Severa Pula, Mallorca

Site Abbreviation: SSP

Inventory Number: 23a

DAMARC — Dating Analyses Index Cards

Material:	H. Bone
Coll. Date:	1979 ?
Lab. No.:	BM-1511
Laboratory Facility:	British Museum
Uncalibrated Dates:	2670 yrs. b.p. 720 ± 60 b.c.
BRCP Calibrated Dates:	

Observations

Part of C. Topp and G. Fernandez series from Formentera.

Site: Ca Na Costa, Formentera

Site Abbreviation: CNC

Inventory Number: 23b

DAMARC — Dating Analyses Index Cards

Material:	Charcoal
Coll. Date:	1968 ?
Lab. No.:	HU-1718
Laboratory Facility:	University of Hamburg
Uncalibrated Dates:	2690 yrs. b.p. 740 ± 60 b.c.
BRCP Calibrated Dates:	

Observations

Date published by German Expedition Sigmund Frey.

Site: Talayot of S'Illot, Mallorca

Site Abbreviation: SILL

Inventory Number: 24

DAMARC — Dating Analyses Index Cards

Material:	Charcoal
Coll. Date:	1980 ?
Lab. No.:	UGRA-123
Laboratory Facility:	University of Granada, Spain
Uncalibrated Dates:	2700 yrs. b.p. 750 ± 170 b.c.
BRCP Calibrated Dates:	

Observations

Part of V. Lull series from Son Fornes.

Site: Settlement of Son Fornes, Mallorca

Site Abbreviation: FORN

Inventory Number: 24a

DAMARC — Dating Analyses Index Cards

Observations

One of the oldest dates marking the introduction of iron at Matge.

Material: Charcoal

Coll. Date: 1969

Lab. No.: QL-11

Laboratory Facility: University of Washington

Uncalibrated Dates: 2700 yrs. b.p. 750 ± 80 b.c.

BRCP Calibrated Dates:

Site: Rock Shelter of Son Matge, Mallorca

Site Abbreviation: ABSM

Inventory Number: 25

DAMARC — Dating Analyses Index Cards

Observations

Part of Torralba series marking Talayotic levels and construction on exterior of the Taula.

Material: Charcoal

Coll. Date: 1979

Lab. No.: QL-1165

Laboratory Facility: University of Washington

Uncalibrated Dates: 2710 yrs. b.p. 760 ± 50 b.c.

BRCP Calibrated Dates:

Site: Taula de Torralba den Salort, Menorca

Site Abbreviation: TT

Inventory Number: 25a

DAMARC — Dating Analyses Index Cards

Material:	Charcoal
Coll. Date:	1969
Lab. No.:	QL-7
Laboratory Facility:	University of Washington
Uncalibrated Dates:	2730 yrs. b.p. 780 ± 100 b.c.
BRCP Calibrated Dates:	

Observations

Oldest date for the introduction of iron at Matge.

Site:	Rock Shelter of Son Matge, Mallorca
Site Abbreviation:	ABSM
Inventory Number:	26

DAMARC — Dating Analyses Index Cards

Material:	Charcoal
Coll. Date:	1962
Lab. No.:	SI-552
Laboratory Facility:	Smithsonian Institution
Uncalibrated Dates:	2765 yrs. b.p. 815 ±120 b.c.
BRCP Calibrated Dates:	

Observations

Talayotic level in Muleta stratigraphy associated with pottery.

Site:	Cave of Son Muleta, Mallorca
Site Abbreviation:	SM
Inventory Number:	27

DAMARC — Dating Analyses Index Cards

Material:	Charcoal
Coll. Date:	1980 ?
Lab. No.:	HAR-2909
Laboratory Facility:	Harwell, England
Uncalibrated Dates:	2770 yrs. b.p. 820 ±100 b.c.
BRCP Calibrated Dates:	

Observations

Analysis for C. Veny on the Naveta of Son Morell, on Menorca.

Site:	Naveta of Son Morell, Menorca
Site Abbreviation:	NSM
Inventory Number:	27a

DAMARC — Dating Analyses Index Cards

Material:	Charcoal
Coll. Date:	1968
Lab. No.:	QL-986
Laboratory Facility:	University of Washington
Uncalibrated Dates:	2820 yrs. b.p. 860 ±50 b.c.
BRCP Calibrated Dates:	

Observations

Date for the ceramic kiln fires at Matge.

Site:	Rock Shelter of Son Matge, Mallorca
Site Abbreviation:	ABSM
Inventory Number:	28

DAMARC — Dating Analyses Index Cards

Material:	Charcoal
Coll. Date:	1979
Lab. No.:	QL-1164
Laboratory Facility:	University of Washington
Uncalibrated Dates:	2830 yrs. b.p. 880 ± 40 b.c.
BRCP Calibrated Dates:	

Observations

Date for the foundations of the Taula on exterior. First dates for what is believed to be a near construction date.

Site: Taula of Torralba den Salort, Menorca

Site Abbreviation: TT

Inventory Number: 29

DAMARC — Dating Analyses Index Cards

Material:	Charcoal
Coll. Date:	1979
Lab. No.:	QL-1089
Laboratory Facility:	University of washington
Uncalibrated Dates:	2840 yrs. b.p. 890 ± 30 b.c.
BRCP Calibrated Dates:	

Observations

Date also dates an early level of the Taula Sample taken from a crevice in bedrock in the NE Annex of the Taula.

Site: Taula of Torralba den Salort, Menorca

Site Abbreviation: TT

Inventory Number: 29a

DAMARC — Dating Analyses Index Cards

Material:	Charcoal
Coll. Date:	1981
Lab. No.:	BM-1697
Laboratory Facility:	University of Washington
Uncalibrated Dates:	2860 yrs. b.p. 910 ± 45 b.c.
BRCP Calibrated Dates:	

Observations

Date for adjacent Naviforme Talayot to the Taula of Torralba.

Site: Taula of Torralba den Salort, Menorca

Site Abbreviation: TT

Inventory Number: 29b

DAMARC — Dating Analyses Index Cards

Material:	Charcoal
Coll. Date:	?
Lab. No.:	GIF-1247
Laboratory Facility:	French Laboratory
Uncalibrated Dates:	2900 yrs. b.p. 950 ± 120 b.c.
BRCP Calibrated Dates:	

Observations

Published G.Lilliu date for Ses Paisses.

Site: Settlement of Ses Paisses, Mallorca

Site Abbreviation: SP

Inventory Number: 30

DAMARC — Dating Analyses Index Cards

Observations

Material: Charcoal

Coll. Date: 1979

Lab. No.: HAR-3413

Laboratory Facility: Harvell, England

Uncalibrated Dates: 2910 yrs. b.p. 960 ±100 b.c.

BRCP Calibrated Dates:

Site: Ferrandell-Oleza Younger Settlement, Ma

Site Abbreviation: SFO-YS

Inventory Number: 30a

DAMARC — Dating Analyses Index Cards

Observations

Material:

Coll. Date:

Lab. No.:

Laboratory Facility: University of Washington

Uncalibrated Dates: 2920 yrs. b.p. 970 ±60 b.c.

BRCP Calibrated Dates:

Site: Settlement of Son Oms, Mallorca

Site Abbreviation: SO

Inventory Number: 31

DAMARC

Dating Analyses Index Cards

Material: Charcoal

Coll. Date: 1967

Lab. No.: Y-1857

Observations

Part of the first Mallorcan series given at the X National Archaeological Congress in Menorca.

Laboratory Facility: Yale University

Uncalibrated Dates: 2960 yrs b.p. 1010 ± 120 b.c.

BRCP Calibrated Dates:

Site: Son Real Figueral, Mallorca

Site Abbreviation: SRF

Inventory Number: 32

DAMARC

Dating Analyses Index Cards

Material: Charcoal

Coll. Date: 1967 ?

Lab. No.: HU-1717

Observations

Part of Geraman series of S. Frey at the Talayotic settlement of S'Illot.

Laboratory Facility: University of Hamburg

Uncalibrated Dates: 2960 yrs. b.p. 1010 ± 110 b.c.

BRCP Calibrated Dates:

Site: Settlement of S'Illot., Mallorca

Site Abbreviation: SILL

Inventory Number: 33

DAMARC — Dating Analyses Index Cards

Material:	Charcaol
Coll. Date:	1979
Lab. No.:	HAR--2908
Observations	Part of Torralba series from the outside of the taula Sanctuary and of Talayotic Age.
Laboratory Facility:	Harwell, England
Uncalibrated Dates:	2970 yrs. b.p. 1020 ± 70 b.c.
BRCP Calibrated Dates:	
Site:	Taula of Torralba den Salort, Menorca
Site Abbreviation:	TT
Inventory Number:	33a

DAMARC — Dating Analyses Index Cards

Material:	Charcoal
Coll. Date:	1967
Lab. No.:	Y-1857
Observations	
Laboratory Facility:	Yale University
Uncalibrated Dates:	3000 yrs. b.p. 1050 ± 120 b.c.
BRCP Calibrated Dates:	
Site:	Son Real Figueral, Mallorca
Site Abbreviation:	SRF
Inventory Number:	34

DAMARC — Dating Analyses Index Cards

Material:	Charcoal
Coll. Date:	1980
Lab. No.:	HAR-?
Laboratory Facility:	Harwell, England
Uncalibrated Dates:	3020 yrs. b.p. 1070 ± 60 b.c.
BRCP Calibrated Dates:	

Observations

Site:	Taula of Torralba den Salort, Menorca
Site Abbreviation:	TT
Inventory Number:	34a

DAMARC — Dating Analyses Index Cards

Material:	Charcoal
Coll. Date:	1980
Lab. No.:	QL-1433
Laboratory Facility:	University of Washington
Uncalibrated Dates:	3030 yrs. b.p. 1080 ± 70 b.c.
BRCP Calibrated Dates:	

Observations

Date conducted on carbonised grain part of Torralba series.

Site:	Taula of Torralba den Salort, Menorca
Site Abbreviation:	TT
Inventory Number:	34b

DAMARC — Dating Analyses Index Cards

Material:	A. Bone
Coll. Date:	1983
Lab. No.:	HAR-3490
Observations	Abandonment level in the Ferrandell-Oleza Old Settlement.
Laboratory Facility:	Harwell, England
Uncalibrated Dates:	3070 yrs. b.p. 1120 ± 50 b.c.
BRCP Calibrated Dates:	

Site: Ferrandell-Oleza Old Settlement, Mallorca

Site Abbreviation: SFO-OS

Inventory Number: 34c

DAMARC — Dating Analyses Index Cards

Material:	Charcoal
Coll. Date:	1967 ?
Lab. No.:	HU-1716
Observations	
Laboratory Facility:	University of Hamburg
Uncalibrated Dates:	3080 yrs. b.p. 1130 ± 75 b.c.
BRCP Calibrated Dates:	

Site: Settlement of S'Illot, Mallorca

Site Abbreviation: SILL

Inventory Number: 35

DAMARC | Dating Analyses Index Cards

Material: A. Bone

Coll. Date: 1983

Lab. No.: BM-1698

Observations

Abondonment date conducted on animal bone at Ferrandell Oleza Old Settlement.

Laboratory Facility: British Museum

Uncalibrated Dates: 3095 yrs. b.p. 1145 ± 70 b.c.

BRCP Calibrated Dates:

Site: Ferrandell-Oleza Old Settlement, Mallorca

Site Abbreviation: SFO-OS

Inventory Number: 35a

DAMARC | Dating Analyses Index Cards

Material: Charcoal

Coll. Date: 1969

Lab. No.: Y-2667

Observations

Date for Talayotic burial levels at Matge.

Laboratory Facility: Yale University

Uncalibrated Dates: 3095 yrs. b.p. 1250 ± 80 b.c.

BRCP Calibrated Dates:

Site: Rock Shelter of Son Matge, Mallorca

Site Abbreviation: ABSM

Inventory Number: 35a

DAMARC — Dating Analyses Index Cards

Material:	A. Bone
Coll. Date:	1983
Lab. No.:	BM-2312
Observations	Animal bone debris from inside Water Channel fill from Ferrandell-Oleza Old Settlement. Occupation level.
Laboratory Facility:	British Museum
Uncalibrated Dates:	3210 yrs. b.p. 1260 ± 80 b.c.
BRCP Calibrated Dates:	

Site: Ferrandell-Oleza Old Settlement, Mallorca

Site Abbreviation: SFO-OS

Inventory Number: 36a

DAMARC — Dating Analyses Index Cards

Material:	Charcoal
Coll. Date:	1975 ?
Lab. No.:	UP-1438
Observations	Date for Rosello Bordoy of the Museum of Mallorca. Earliest Talayotic context at present.
Laboratory Facility:	University of Pennsylvania
Uncalibrated Dates:	3260 yrs. b.p. 1310 ± 60 b.c.
BRCP Calibrated Dates:	

Site: Settlement of Son Severa Pula, Mallorca

Site Abbreviation: SP

Inventory Number: 37

DAMARC — Dating Analyses Index Cards

Material:	H. Bone
Coll. Date:	1974 ?
Lab. No.:	BM-1677
Laboratory Facility:	British Museum
Uncalibrated Dates:	3270 yrs. b.p. 1320 ± 80 b.c.
BRCP Calibrated Dates:	

Observations

Part of C. Topp and G. Fernadez series
Oldest date associated with megalithic tomb.

Site: Ca Na Costa, Formentera
Site Abbreviation: CNC
Inventory Number: 37a

DAMARC — Dating Analyses Index Cards

Material:	Hh. Carbon
Coll. Date:	1971
Lab. No.:	QL-5
Laboratory Facility:	University of Washington
Uncalibrated Dates:	3350 yrs. b.p. 1400 ± 60 b.c.
BRCP Calibrated Dates:	

Observations

Last Pretalayotic hearth levels at Matge.
Test done on hearth carbonates from
Kitchen fires. Marlks the abandonment
of the site for living in the Pretalayotic.

Site: Rock Shelter of Son Matge, Mallorca
Site Abbreviation: ABSM
Inventory Number: 38

DAMARC — Dating Analyses Index Cards

Material:	Hh. Carbon
Coll. Date:	1982
Lab. No.:	BM-1995
Laboratory Facility:	British Museum
Uncalibrated Dates:	3380 yrs. b.p. 1430 ± 50 b.c.
BRCP Calibrated Dates:	

Observations

Date from a hearth fire, burnt bone.

Site:	Rock Shelter of Son Matge, Mallorca
Site Abbreviation:	ABSM
Inventory Number:	38a

DAMARC — Dating Analyses Index Cards

Material:	Hh. Carbon
Coll. Date:	1970
Lab. No.:	QL-5a
Laboratory Facility:	University of Washington
Uncalibrated Dates:	3420 yrs. b.p. 1470 ± 100 b.c.
BRCP Calibrated Dates:	

Observations

Date from a hearth zone on fire carbonates from the last habitational level at Matge.

Site:	Rock Shelter of Son Matge, Mallorca
Site Abbreviation:	ABSM
Inventory Number:	33

DAMARC **Dating Analyses Index Cards**

Material:	H. Bone
Coll. Date:	1966
Lab. No.:	Y-1856
Laboratory Facility:	Yale University
Uncalibrated Dates:	3470 yrs. b.p. 1520 ± 80 b.c.
BRCP Calibrated Dates:	

Observations

Date on human bone from secondary burials. First initial series.

Site:	Cave of Son Marroig, Mallorca
Site Abbreviation:	SMRG
Inventory Number:	40

DAMARC **Dating Analyses Index Cards**

Material:	Hb. Carbon
Coll. Date:	1971
Lab. No.:	CISIC-179
Laboratory Facility:	University of Madrid
Uncalibrated Dates:	3480 yrs. b.p. 1530 ± 80 b.c.
BRCP Calibrated Dates:	

Observations

Part of Madrid series for correlation with rest of the Matge series from the University of Washington.

Site:	Rock Shelter of Son Matge, Mallorca
Site Abbreviation:	ABSM
Inventory Number:	41

DAMARC — Dating Analyses Index Cards

Material: A. Bone

Coll. Date: 1983

Lab. No.: QL-1859

Observations
Date on animal bone from the fill of a ancient terrace AT1-AT2. Represents occupation or construction phase of the terrace.

Laboratory Facility: University of Washington

Uncalibrated Dates: 3490 yrs. b.p. 1540 80 b.c.

BRCP Calibrated Dates:

Site: Ferrandell-Oleza Old Settlement, Mallorca

Site Abbreviation: SFO-OS

Inventory Number: 41a

DAMARC — Dating Analyses Index Cards

Material: Hh. Carbon

Coll. Date: 1978

Lab. No.: CISIC-179

Observations
Date from hearth zone. charcoal and burnt animal bone. Part of Madrid series.

Laboratory Facility: University of Madrid

Uncalibrated Dates: 3620 yrs. b.p. 1620 ± 80 b.c.

BRCP Calibrated Dates:

Site: Rock Shelter of Son Matge, Mallorca

Site Abbreviation: ABSM

Inventory Number: 42

DAMARC

Dating Analyses Index Cards

Material: Hh. Carbon

Coll. Date: 1971

Lab. No.: QL-24

Observations

Date from hearth sequence at Matge.

Laboratory Facility: University of Washington

Uncalibrated Dates : 3670 yrs. b.p. 1720 ± 100 b.c.

BRCP Calibrated Dates:

Site: Rock Shelter of Son Matge, Mallorca

Site Abbreviation: ABSM

Inventory Number: 43

DAMARC

Dating Analyses Index Cards

Material: Charcoal

Coll. Date: 1982

Lab. No.: QL-1592

Observations

Date from West Old Settlement fire in the refuse area.

Laboratory Facility: University of Washington

Uncalibrated Dates : 3700 yrs. b.p. 1750 ± 30 b.c.

BRCP Calibrated Dates:

Site: Ferrandell-Oleza Old Settlement, Mallorca

Site Abbreviation: SFO-OS

Inventory Number: 43a

DAMARC — **Dating Analyses Index Cards**

Material:	Hh.Carbon
Coll. Date:	1983
Lab. No.:	BM-1981
Laboratory Facility:	British Museum
Uncalibrated Dates:	3720 yrs. b.p. 1770 ± 35 b.c.
BRCP Calibrated Dates:	

Observations

Date from hearth fire in refuse area of West Old Settlement.

Site: Ferrandell-Oleza Old settlement, Mallorca

Site Abbreviation: SFO-OS

Inventory Number: 43b

DAMARC — **Dating Analyses Index Cards**

Material:	Charcoal
Coll. Date:	1972 ?
Lab. No.:	I-5515
Laboratory Facility:	Iostopes Inc, USA
Uncalibrated Dates:	3750 yrs. b.p. 1800 ± 120 b.c.
BRCP Calibrated Dates:	

Observations

Date on charcoal from hearth fire at Coxters.

Site: Ca Na Cotxera, Mallorca

Site Abbreviation: CX

Inventory Number: 44

DAMARC — Dating Analyses Index Cards

Material: Charcoal

Coll. Date: 1967

Lab. No.: Y-1789

Observations

Date on charcoal associated with a burial containing the first Bell Beaker ware from the rock shelter of Muertos.

Laboratory Facility: Yale University

Uncalibrated Dates: 3790 yrs. b.p. 1840 ± 80 b.c.

BRCP Calibrated Dates:

Site: Rock Shelter of Muertos Gallard, Mallorca

Site Abbreviation: AMG

Inventory Number: 45

DAMARC — Dating Analyses Index Cards

Material: Charcoal

Coll. Date: 1983

Lab. No.: QL-1636

Observations

Charcoal from a fire at the base of the North Compound Wall Entrance and associated with Beaker Ware.

Laboratory Facility: University of Washington

Uncalibrated Dates: 3790 yrs.b.p. 1840 ± 90 b.c.

BRCP Calibrated Dates:

Site: Ferrandell-Oleza Old Settlement, Mallorca

Site Abbreviation: SFO-OS

Inventory Number: 45a

DAMARC — Dating Analyses Index Cards

Material: Charcoal

Coll. Date: 1968

Lab. No.: Y-2359

Observations

Charcoal from a hearth associate with Beaker pottery at Matge.

Laboratory Facility: Yale University

Uncalibrated Dates: 3820 yrs.b.p. 1870 ± 120 b.c.

BRCP Calibrated Dates:

Site: Rock Shelter of Son Matge, Mallorca

Site Abbreviation: ABSM

Inventory Number: 46

DAMARC — Dating Analyses Index Cards

Material: Charcoal

Coll. Date: 1962

Lab. No.: Y-2359a

Observations

Date on charcoal from a habitation level at Muleta associated with Beaker ware undecorated typology.

Laboratory Facility: Yale University

Uncalibrated Dates: 3910 yrs. b.p. 1960 ± 120 b.c.

BRCP Calibrated Dates:

Site: Cave of Son Muleta, Mallorca

Site Abbreviation: SM

Inventory Number: 47

DAMARC — Dating Analyses Index Cards

Material:	Charcoal
Coll. Date:	1980
Lab. No.:	BM-1843
Laboratory Facility:	British Museum
Uncalibrated Dates:	3950 yrs. b.p. 2000 ± 65 b.c.
BRCP Calibrated Dates:	

Observations

Charcoal from a hearth fire at the base of the East Compound Wall associated with Beaker Ware.

Site:	Ferrandell-Oleza Old Settlement, Mallorca
Site Abbreviation:	SFO-OS
Inventory Number:	47a

DAMARC — Dating Analyses Index Cards

Material:	Hh.Carbon
Coll. Date:	1969
Lab. No.:	QL-5b
Laboratory Facility:	University of Washington
Uncalibrated Dates:	3970 yrs. b.p. 2020 ± 100 b.c.
BRCP Calibrated Dates:	

Observations

Date from hearth fire. Part of Matge sequence.

Site:	Rock Shelter of Son Matge, Mallorca
Site Abbreviation:	ABSM
Inventory Number:	48

DAMARC — Dating Analyses Index Cards

Material:	Hh.Carbon
Coll. Date:	1969
Lab. No.:	CSIC-178
Observations	Date from the hearth fire sequence at Matge.
Laboratory Facility:	University of Madrid
Uncalibrated Dates:	3980 yrs. b.p. 2030 ± 170 b.c.
BRCP Calibrated Dates:	
Site:	Rock Shelter of Son Matge, Mallorca
Site Abbreviation:	ABSM
Inventory Number:	49

DAMARC — Dating Analyses Index Cards

Material:	Hh.Carbon
Coll. Date:	1970
Lab. No.:	QL-23
Observations	Date from the hearth fire sequence at Matge.
Laboratory Facility:	University of Washington
Uncalibrated Dates:	4020 yrs. b.p. 2070 ± 50 b.c.
BRCP Calibrated Dates:	
Site:	Rock Shelter of Son Matge, Mallorca
Site Abbreviation:	ABSM
Inventory Number:	50

DAMARC — **Dating Analyses Index Cards**

Material: M. Bone

Coll. Date: 1981

Lab. No.: BM-1408

Observations

Myotragus bone from habittional fires at Matge. Survival of M. balearicus until this date. Associated with domestic animals.

Laboratory Facility: British Museum

Uncalibrated Dates: 4093 yrs. b.p. 2143 ±392 b.c.

BRCP Calibrated Dates:

Site: Rock Shelter of Son Matge, Mallorca

Site Abbreviation: ABSM

Inventory Number: 50a

DAMARC — **Dating Analyses Index Cards**

Material: Hh.Carbon

Coll. Date: 1978

Lab. No.: QL-988

Observations

Date for oldest pottery level and earliest introduction of domesticated animals at Matge.

Laboratory Facility: University of Washington

Uncalibrated Dates: 4650 yrs. b.p. 2700 120 b.c.

BRCP Calibrated Dates:

Site: Rock Shelter of Son Matge, Mallorca

Site Abbreviation: ABSM

Inventory Number: 51

DAMARC — **Dating Analyses Index Cards**

Material:	Carcoal
Coll. Date:	1983
Lab. No.:	BM-1994
Observations	Earliest date for pottery, correlates with Matge.
Laboratory Facility:	British Museum
Uncalibrated Dates :	4740 yrs. b.p. 2810 ± 50 b.c.
BRCP Calibrated Dates:	
Site:	Rock Shelter of Muertos Gallard, Mallorca
Site Abbreviation:	AMG
Inventory Number :	51a

DAMARC — **Dating Analyses Index Cards**

Material:	Hh.Carbon
Coll. Date:	1972
Lab. No.:	I-5516
Observations	Oldest habitational level at Matge.
Laboratory Facility:	Isotopes Inc., USA
Uncalibrated Dates :	5750 yrs. b.p. 3800 ± 115 b.c.
BRCP Calibrated Dates:	
Site:	Rock Shelter of Son Matge, Mallorca
Site Abbreviation:	ABSM
Inventory Number :	52

DAMARC — Dating Analyses Index Cards

Material:	Coprolites
Coll. Date:	1978
Lab. No.:	CSIC-176
Laboratory Facility:	University of Madrid
Uncalibrated Dates:	5820 yrs. b.p. 3870 ± 360 b.c.
BRCP Calibrated Date:	

Observations

Date of M. balearicus coprolitest from a level (Stratum 333) at Matge.

Site:	Rock Shelter of Son Matge, Mallorca
Site Abbreviation:	ABSM
Inventory Number:	53

DAMARC — Dating Analyses Index Cards

Material:	H. bone
Coll. Date:	1963
Lab. No.:	KBN-640b
Laboratory Facility:	Karman BioNuclear
Uncalibrated Dates:	5935 yrs. b.p. 3985 ±10? b.c.
BRCP Calibrated Dates:	

Observations

Human bone associated with M. balearicus from Matge. Oldest date for the remains of man.

Site:	Cave of Son Muleta, Mallorca
Site Abbreviation:	SM
Inventory Number:	54

DAMARC — **Dating Analyses Index Cards**

Material: M.Bone

Coll. Date: 1970

Lab. No.: QL-29

Observations

Oldest dated human activity level at Matge. M.balearicus bone with butcher marks and trimmed horn cores.

Laboratory Facility: University of Washington

Uncalibrated Dates: 6680 yrs. b.p. 4730 ± 120 b.c.

BRCP Calibrated Dates:

Site: Rock Shelter of Son Matge, Mallorca

Site Abbreviation: ABSM

Inventory Number: 55

DAMARC — **Dating Analyses Index Cards**

Material: M. Bone

Coll. Date: 1963

Lab. No.: KBN-460c

Observations

Date on youngest M.balearicus bone from Muleta associated with man.

Laboratory Facility: Karman BioNuclear

Uncalibrated Dates: 7135 yrs. b.p. 5185 ± 80 b.c.

BRCP Calibrated Dates:

Site: Cave of Son Muleta, Mallorca

Site Abbreviation: SM

Inventory Number: 56

DAMARC — **Dating Analyses Index Cards**

Material:	M.Bone.
Coll. Date:	1974
Lab. No.:	UCLA-1704j
Laboratory Facility:	University of California at LA
Uncalibrated Dates:	7300 yrs b.p. 6350 ± 100 b.c.
BRCP Calibrated Dates:	

Observations

Part of correlation series with RAA dates at Muleta

Site:	Cave of Son Muleta, Mallorca
Site Abbreviation:	SM
Inventory Number:	56a

DAMARC — **Dating Analyses Index Cards**

Material:	M. Bone
Coll. Date:	1983
Lab. No.:	HAR-2934
Laboratory Facility:	Harwell, England
Uncalibrated Dates:	16,500 yrs. b.p. 15,500 2300 b.c.
BRCP Calibrated Dates:	

Observations

Part of A. Rae series for UT dating of M. balearicus bone from Muleta.

Site:	Cave of Son Muleta, Mallorca
Site Abbreviation:	SM
Inventory Number:	56b

DAMARC — Dating Analyses Index Cards

Material:	M. Bone
Coll. Date:	1973
Lab. No.:	UCLA-1704c
Observations	Part of RAA dating survey at Muleta
Laboratory Facility:	University of California at LA
Uncalibrated Dates:	8570 yrs. b.p. 6620 ± b.c.
BRCP Calibrated Dates:	

Site:	Cave of Son Muleta, Mallorca
Site Abbreviation:	SM
Inventory Number:	57

DAMARC — Dating Analyses Index Cards

Material:	M. Bone
Coll. Date:	1973
Lab. No.:	UCLA-1704f
Observations	Part of RAA series at Muleta
Laboratory Facility:	University of California at LA
Uncalibrated Dates:	8800 yrs. b.p. 6850 ±400 b.c.
BRCP Calibrated Dates:	

Site:	Cave of Son Muleta, Mallorca
Site Abbreviation:	SM
Inventory Number:	57a

DAMARC — Dating Analyses Index Cards

Material: M. Bone

Coll. Date: 1983

Lab. No.: HAR-2909

Observations

Part of A. Rae UT dating series of M. balearicus bone at Muleta.

Laboratory Facility: Harwell, England

Uncalibrated Dates: 16,400 yrs. b.p. 14,500 ± 1200 b.c.

BRCP Calibrated Dates:

Site: Cave of Son Muleta, Mallorca

Site Abbreviation: SM

Inventory Number: 57b

DAMARC — Dating Analyses Index Cards

Material: M. Bone

Coll. Date: 1983

Lab. No.: HAR-2932

Observations

Part of A. Rae UT dating of M. balearicus bone at Muleta

Laboratory Facility: Harwell, England

Uncalibrated Dates: 19,700 yrs b.p. 17,750 ± 1110 b.c.

BRCP Calibrated Dates:

Site: Cave of Son Muleta, Mallorca

Site Abbreviation: SM

Inventory Number: 57c

DAMARC — Dating Analyses Index Cards

±

Material: M.Bone

Coll. Date: 1962

Lab. No.: KBN-640 b

Observations

First date on M. Balearicus bone at Muleta. Sample 38.5 grms large statistical margin of error.

Laboratory Facility: Karman BioNuclear

Uncalibrated Dates : 8448 yrs. b.p. 6498 1180 b.c.

BRCP Calibrated Dates:

Site: Cave of Son Muleta, Mallorca

Site Abbreviation: SM

Inventory Number: 58

DAMARC — Dating Analyses Index Cards

Material: H. Bone

Coll. Date: 1963

Lab. No.: KBN-640c

Observations

Date on human bone. Firts sample to small for accurate date. Rerun made.

Laboratory Facility: Karman BioNucleur

Uncalibrated Dates : 10,685 yrs b.p. 8735 ±3517 b.c.

BRCP Calibrated Dates:

Site: Cave of Son Muleta, Mallorca

Site Abbreviation: SM

Inventory Number: 59

DAMARC

Dating Analyses Index Cards

±

Material: M. Bone

Coll. Date: 1969

Lab. No.: SI0f

Observations

Laboratory Facility: ș Scripps Institute ıtion

Uncalibrated Dates : Pending

BRCP Calibrated Dates:

Site: Cave of Son Muleta, Mallorca

Site Abbreviation: SM

Inventory Number: 60

DAMARC

Dating Analyses Index Cards

Material: M. Bone

Coll. Date: 1973

Lab. No.: SI-654a

Observations

First of the RAA dates from Scripps for correlation with radiocarbon series.

Laboratory Facility: Smithsonian Institution

Uncalibrated Dates : 14,000 yrs b.p. 12,050 350 b.c.

BRCP Calibrated Dates:

Site: Cave of Son Muleta, Mallorca

Site Abbreviation: SM

Inventory Number: 61

DAMARC — Dating Analyses Index Cards

Material: M. Bone

Coll. Date: 1973

Lab. No.: SI-654b

Observations: Part of RAA series at Scripps

Laboratory Facility: Smithsonian Institution

Uncalibrated Dates: 14,465 yrs b.p. 12,515± 315 b.c.

BRCP Calibrated Dates:

Site: Cave of Son Muleta, Mallorca

Site Abbreviation: SM

Inventory Number: 62

DAMARC — Dating Analyses Index Cards

Material: M. Bone

Coll. Date: 1973

Lab. No.: UCLA-170b

Observations: Radiocarbon date for correlation with RAA series at Muleta.

Laboratory Facility: University of California at LA

Uncalibrated Dates: 14,650 yrs b.p 12,700 850 b.c.

BRCP Calibrated Dates:

Site: Cave of Son Muleta, Mallorca

Site Abbreviation: SM

Inventory Number: 63

DAMARC — **Dating Analyses Index Cards**

Observations

Part of radiocarbon series from Muleta on M. balearicus bone correlation with Scripps and UCLA

Material: M. Bone

Coll. Date: 1978

Lab. No.: SI-646

Laboratory Facility: Smithsonian Institution

Uncalibrated Dates : 15,885 yrs b.p. 13,985 ± 655 b.c.

BRCP Calibrated Dates:

Site: Cave of Son Muleta, Mallorca

Site Abbreviation: SM

Inventory Number: 64

DAMARC — **Dating Analyses Index Cards**

Observations

Part of the A. Rae 14C date associated with UT series.

Material: M. Bone

Coll. Date: 1973

Lab. No.: HAR-2933

Laboratory Facility: Harwell, England

Uncalibrated Dates : 23,500 yrs b.p. 21,550 1600 b.c.

BRCP Calibrated Dates:

Site: Cave of Son Muleta, Mallorca

Site Abbreviation: SM

Inventory Number: 64a

DAMARC

Dating Analyses Index Cards

Material: M. Bone

Coll. Date: 1973

Lab. No.: SI- 648

Observations

Part of radiocarbon series from Muleta on M. balearicus bone for possible correlation with RAA series and UCLA 14C dates

Laboratory Facility: Smithsonian Institution

Uncalibrated Dates: 16,335 yrs b.p. 14,385 ± 415 b.c.

BRCP Calibrated Dates:

Site: Cave of Son Muleta, Mallorca

Site Abbreviation: SM

Inventory Number: 65

DAMARC

Dating Analyses Index Cards

Material: M. Bone

Coll. Date: 1973

Lab. No.: UCLA-1704d

Observations

Part of the RAA series at Scripps from Muleta, radiocarbon date.

Laboratory Facility: University of California at LA

Uncalibrated Dates: 16,850 yrs b.p. 14,900 200 b.c.

BRCP Calibrated Dates:

Site: Cave of Son Muleta, Mallorca

Site Abbreviation: SM

Inventory Number: 66

DAMARC — Dating Analyses Index Cards

Material:	M. Bone
Coll. Date:	1984
Lab. No.:	HAR-2931
Observations	Part of A. Rae UT Survey
Laboratory Facility:	HARWELL, ENGLAND
Uncalibrated Dates:	14,200 yrs. b.p. 12,250 ± 1100 b.c.
BRCP Calibrated Dates:	
Site:	Cave of Son Muleta, Mallorca
Site Abbreviation:	SM
Inventory Number:	66a

DAMARC — Dating Analyses Index Cards

Material:	M.Bone
Coll. Date:	1968
Lab. No.:	SI-649
Observations	Part of the Smithsonian Series for cross-correlation with RAA survey
Laboratory Facility:	Smithsonian Institute
Uncalibrated Dates:	18,100 yrs. b p 16,150 600 b.c.
BRCP Calibrated Dates:	
Site:	Cave of Son Muleta, Mallorca
Site Abbreviation:	SM
Inventory Number:	67

DAMARC

Dating Analyses Index Cards

±

Material: M.Bone

Coll. Date: 1972

Lab. No.: SIO-1i

Observations

Laboratory Facility: Scripps Institute

Uncalibrated Dates : Pending

BRCP Calibrated Dates:

Site: Cave of Son Muleta, Mallorca

Site Abbreviation: SM

Inventory Number: 68

DAMARC

Dating Analyses Index Cards

Material: M.Bone

Coll. Date: 1968

Lab. No.: SI-650

Observations

Laboratory Facility: Smithsonian Institution

Uncalibrated Dates : 18,375 yrs b.p. 16,805 ± 555 b.c.

BRCP Calibrated Dates:

Site: Cave of Son Muleta, Mallorca

Site Abbreviation: SM

Inventory Number: 69

DAMARC — **Dating Analyses Index Cards**

±

Material: M.Bone

Coll. Date: 1974

Lab. No.: UCLA-1704e

Observations

Laboratory Facility: University of California at LA

Uncalibrated Dates : 18,900 yrs b.p. 17,030 200 b.c.

BRCP Calibrated Dates:

Site: Cave of Son Muleta, Mallorca

Site Abbreviation: SM

Inventory Number: 70

DAMARC — **Dating Analyses Index Cards**

Material: M.Bone

Coll. Date: 1968

Lab. No.: SI-647

Observations

Laboratory Facility: Smithsonian Institute

Uncalibrated Dates : 23,800 yrs b.p. 21,850 480 b.c

BRCP Calibrated Dates:

Site: Cave of Son Muleta, Mallorca

Site Abbreviation: SM

Inventory Number: 71

DAMARC | **Dating Analyses Index Cards**

Material:	M.Bone
Coll. Date:	1983
Lab. No.:	HAR-2912
Observations	
Laboratory Facility:	Harwell, England
Uncalibrated Dates:	23,200 yrs b.p. 21,250 ± 2300 b.c.
BRCP Calibrated Dates:	
Site:	Cave of Son Muleta, Mallorca
Site Abbreviation:	SM
Inventory Number:	71a

DAMARC | **Dating Analyses Index Cards**

Material:	M.Bone
Coll. Date:	1974
Lab. No.:	SIO-1c
Observations	
Laboratory Facility:	Scripps Institute
Uncalibrated Dates:	26,000 yrs b.p. 25,050 ± 600 b.c.
BRCP Calibrated Dates:	
Site:	Cave of Son Muleta, Mallorca
Site Abbreviation:	SM
Inventory Number:	72

DAMARC — Dating Analyses Index Cards

Material: M.Bone

Coll. Date: 1983

Lab. No.: HAR-2912

Observations

Laboratory Facility: Harwell, England

Uncalibrated Dates: 25,800 yrs b.p. 23,850 ± 1300 b.c.

BRCP Calibrated Dates:

Site: Cave of Son Muleta, Mallorca

Site Abbreviation: SM

Inventory Number: 72a

DAMARC — Dating Analyses Index Cards

Material: M.Bone

Coll. Date: 1974

Lab. No.: UCLA-1704c

Observations

Laboratory Facility: University of California at LA

Uncalibrated Dates: 28,600 yrs b.p. 26,650 ± 600 b.c.

BRCP Calibrated Dates:

Site: Cave of Son Muleta, Mallorca

Site Abbreviation: SM

Inventory Number: 73

DAMARC | Dating Analyses Index Cards

Material:	M.Bone
Coll. Date:	1968
Lab. No.:	SIO-1g
Observations	
Laboratory Facility:	Scripps Institute
Uncalibrated Dates:	32,000 yrs b.p. 30,050 ± 715 b.c.
BRCP Calibrated Dates:	

Site:	Cave of Son Muleta, Mallorca
Site Abbreviation:	SM
Inventory Number:	74

DAMARC | Dating Analyses Index Cards

Material:	M.Bone
Coll. Date:	1974
Lab. No.:	SIO-1h
Observations	
Laboratory Facility:	Scripps Institute
Uncalibrated Dates:	42,000 yrs b.p. 40,050 ±1000 b.c.
BRCP Calibrated Dates:	

Site:	Cave of Son Muleta, Mallorca
Site Abbreviation:	SM
Inventory Number:	74a

DAMARC — **Dating Analyses Index Cards**

Material:	M. Bone
Coll. Date:	1983
Lab. No.:	HAR-2911
Laboratory Facility:	Harwell, England
Uncalibrated Dates :	44,800 yrs b.p. 42,850 ± 1600 b.c.
BRCP Calibrated Dates:	

Observations

Site:	Cave of Son Muleta, Mallorca
Site Abbreviation:	SM
Inventory Number:	74b

DAMARC — **Dating Analyses Index Cards**

Material:	Stalagmite
Coll. Date:	1968
Lab. No.:	QL-127
Laboratory Facility:	University of Washington
Uncalibrated Dates :	48,000 yrs b.p. exterior
BRCP Calibrated Dates:	

Observations

Carbonate dating on stalagmite formation

Site:	
Site Abbreviation:	
Inventory Number:	

DAMARC — Dating Analyses Index Cards

Material:	Stalagmite
Coll. Date:	1968
Lab. No.:	1968
Laboratory Facility:	University of Washington
Uncalibrated Dates:	58,000 yrs b.p. interior
BRCP Calibrated Dates:	

Observations: Reading on interior of stalagmite core

Site: Cave of Son Muleta, Mallorca

Site Abbreviation: SM

Inventory Number: 76

DAMARC — Dating Analyses Index Cards

Material:	Calcite
Coll. Date:	1968
Lab. No.:	Y-1f
Laboratory Facility:	Yale University
Uncalibrated Dates:	80,000 yrs b.p.
BRCP Calibrated Dates:	

Observations: Date on flowstone calcite floor in 'X' sector.

Site: Cave of Son Muleta, Mallorca

Site Abbreviation: SM

Inventory Number: 77

Notes on the Analysis of Bronze Samples from the Central Enclosure of the Rock Shelter of Son Matge

<<<<<<<<<<<<<<<<<<<<<<<<<<<<<<<<<<<<<<<<<<<<>>

The results of electron microscope analysis conducted on five samples (copper and bronze), the first of a series in a programme of 200 from prehistoric contexts in the Balearic Islands, are listed below. The five samples originate from three different cultural periods. Three of the samples are from the Pretalayotic Period, circa 1800 b.c. to circa 1400 b.c., while the Talayotic Period, circa 1400 b.c. to circa 800 b.c., and the Post Talayotiic Period circa 800 b.c. to circa 123 B.C. are represented by one sample each. The anlayses were carried out by Dr. Peter J. Northover of the Department of Metallurgy, University of Oxford.

Sample 1. Inventory Number ABSM-EMA 001 (WW1)

The sample was taken from the green encrustration from the inside of a fragment belonging to a ceramic crucible (undecorated). Examination under the electron microscope proved this to be mainly copper oxides with copper corrosion products.

Comments: Pretalayotic Origin (EBP), Stratum 11.

Sample 2. Inventory Number ABSM-EMA 002 (WW2)

The sample was taken from the green encrustration from the inside of a fragment belonging to a ceramic crucible (decorated with Beaker motifs). Examination under the electron microscope proved this to be mainly copper oxides with copper corrosion products.

Comments: Pretalayotic Origin (EBP), Stratum 12.

Sample 3. Inventory Number ABSM-EMA 003 (WW3)

A sample was taken from the point of an awl. This point proved by a semiquantitative analysis to be approximately a 10% tin bronze with a low level of impurities, but numerous small oxide and slag inclusions. It has a fully recrystallised equiaxed grain structure with annealing twins and some slight subsequent deformation. The grain size is 15-20 um

Comments: Pretalayotic Origin (LBP), Stratum 10.

Sample 4. Inventory Number ABSM-EMA 004 (WW4)

Part of a bronze bead was examined and found to be completely corroded. Analysis gave a considerable response from tin. As tin contents are enhanced by corrosion, it is difficult to suggest the original value, but it was probably in the order of 5-10%.

Comments: Post Talayotic Origin (PT), Stratum 2.

Sample 5. Inventory Number ABSM-EMA005 (WW5)

A sample was taken from a pin. A semiquantitative analysis indicated a 6-7% tin bronze with a low level of impurities, but with oxide and slag inclusions. The effects of etching suggests that this bronze is purer than Sample 3. The microstructure is equiaxed and fully recrystallised with annealing twins but no further disformation. Grain size is 40 um. There is very extensive intergranular corrosion attack.

Comments: Talayotic Origin (TP), Stratum 8.

Conclusions

<<<<<<<<<o>>>>>>>>>

While it is premature to draw any definite conclusions as to the source of the bronze from which the Matge specimens were manufactured, it does appear that there are some compositional differences in the tin content which should become more significant when a large series of samples have been analysed. The fact that the crucible samples analysed show copper oxides unalloyed with tin or arsenic may be significant of their age, but it has been noted by Northover that this seems to be a normal occurrence with Beaker crucibles, as well as other criucibles of the same age. Whether or not this characteristic of the residue from the inside of the crucible being copper, rather than bronze, will hold true of all Balearic Beaker crucible sherds cannot be ascertained until a large series of EBP artefacts is analysed. The LBP analysis ,**Sample 3**, is interesting in that it suggests that a true bronze was in use in the Balearic circa 1700 b.c. to circa 1400 b.c., while **Sample 5** from the Talayotic contexts at Matge, circa 1400 b.c. to circa 800 b.c., shows a true tin bronze with a low level of impurities, purer than **Sample 3**. Furthermore, the absolute age of the context from which this piece originates can be determined by radiocarbon dating 1250 b.c. ± 100 yrs. (**ABSM 36**)(**Y-2667**), thus suggesting that the Talayotic bronze was a very refined product. **Sample 4** which is from the Post Talayotic quicklime burial contexts, circa 700 b.c. to 123 B.C., appears also to have had a high tin content, thought the highly corrosive nature of the lime in which the bronze artefacts are found seems to have been a factor in its preservation.

On the whole, the author feels that these first analytical results are encouraging, and that additional analyses should be of further value in determining the composition of the copper and tin bronze of the different periods, as well as eventually giving clues to sources of the ores.

Please note that a full-size version of this page is available to download from www.barpublishing.com/additional-downloads.html. The original foldout has been reduced in size to match the A4 format of this book, the image is therefore not as clear as the original foldout. Please refer to the original foldout via the download for the original content.

	No.	PROVENANCE		cu	Sn	Sb	Pb	Co	Ni	Fe	Ag	Au	Zn	Bi	As	Incs.	State
DAMARC	1	ABSM/CA31-32/L 120 cms.	Hair piece	87.18	10.12	0.00	0.05	ti	0.08	0.00	0.02	0.00	0.02 +	0.02	0.08	------	
DAMARC	2	ABSM/CA31-32/L 120 cms.	Bronze needle	89.47	8.81	0.09	1.31	0.00	0.05	ti	0.05	ti	0.02 +	0.02 +	0.18	------	
DAMARC	3	ABSM/CA31-32/L 120 cms.	Bronze needle	89.75	9.85	0.00	0.08	0.01	0.03	0.02	0.07	ti	ti	0.01	0.18	------	
DAMARC	4	ABSM/CA24/L Talayotic	Bronze needle	90.22	9.52	0.00	0.06	ti	0.02	0.03	0.07	0.00	0.03 +	0.03	0.15	------	
DAMARC	5*	ABSM/CA24/L Talayotic	Bronze Chest	71.60	27.60	0.00	0.17	0.03	0.10	0.07	0.00	0.00	0.07	0.02	0.34	------	*Corrosion product only.
DAMARC	6	ABSM/CA42/L	Handle of awl	99.63	0.00	0.01	0.03	0.01	0.04	0.00	0.09	0.04	ti	0.03	0.12	------	
DAMARC	7	ABSM/CA42/L	Point of awl	91.71	7.68	0.40	ti	ti	0.08	ti	0.06	0.00	0.00	ti	0.07	------	
DAMARC	8*	ABSM/CA26-27/L Rubble Fill of Talayotic Wall.	Bronze frac.	99.58	0.00	0.00	0.06	0.00	0.02	0.00	0.08	0.00	0.00	0.00	0.26	------	*Corrosion product only.
DAMARC	9	ABSM/CA34/L 40 cms.	Bronze horn	93.89	5.66	0.00	0.09	0.01	0.07	0.01	0.05	0.15	0.03 +	0.00	0.04	------	
DAMARC	10	ABSM/CA37/L 40 cms.	Hair piece	88.37	10.64	0.05	0.66	ti	0.05	ti	0.05	0.00	0.00	ti	0.18	------	
DAMARC	11*	ABSM/CA35/L 40 cms.Post Talayotic.	Bronze bead	99.59	0.00	0.00	0.26	0.00	0.00	0.02	0.02	0.04	0.00	0.00	0.07	------	*Corrosion product only.
DAMARC	12	ABSM/CA /L Post Talayotic.	Bronze nail	88.55	10.10	0.04	1.11	0.00	0.04	0.00	0.03	0.00	ti	ti	0.13	------	
DAMARC	13	ABSM/CA37/'P'Rock Post Talayotic.	Tratinabulum	92.27	7.14	0.03	0.30	ti	0.08	0.01	0.06	ti	0.00	0.01 +	0.10	------	
DAMARC	14	ABSM/CA26-27/L 100 cms.	Bronze pieze	88.50	11.27	0.00	0.00	0.01	ti	0.01	0.04	0.00	ti	0.04	0.13	------	
DAMARC	15	ABSM/CA26-27/L 100 cms.	Tratinabulum	91.43	7.70	0.04	0.60	0.01	0.07	0.02	0.06	0.00	0.04	0.02 +	0.10	------	
DAMARC	16	ABSM/CA11-16/L	Tratinabulum	94.22	5.31	0.04	0.05	0.02	0.08	ti	0.13	0.02 +	0.00	0.01 +	0.12	------	
DAMARC	17	ABSM/CA 30/L	Bronce nail	95.31	4.21	0.01	0.13	0.01	0.10	0.08	0.05	0.02 +	0.00	0.01 +	0.07	------	
DAMARC	18	ABSM/CA 30/L	Bronze frac.	90.04	8.95	0.03	0.50	0.01	0.08	0.08 +	0.03	ti	0.06 +	0.00	0.22	------	
DAMARC	19	ABSM/CA 30/L	Tratinabulum	90.46	8.91	0.05	0.20	0.01 +	0.08	0.02	0.06	0.02 +	0.03 +	0.01 +	0.15	------	
DAMARC	20*	ABSM/CA 23/L 80-90 cms.	Frag.	52.75	46.69	0.00	0.00	0.00	0.03	0.03	0.04	0.00	0.06	0.00	0.40	------	*Corrosion product only.
DAMARC	21	ABSM/CA 22/L	Bronze spiral	90.15	8.69	0.09	0.75	0.01	0.06	0.00	ti	0.07 +	0.00	0.01 +	0.17	------	
DAMARC	22*	ABSM/CA 23/ L 80-90 cms.	Piece of coil	99.54	0.04	0.00	0.09	0.00	0.04	0.01	0.02	0.24	0.02	0.00	0.00	------	*Corrosion product only.
DAMARC	23	ABSM/CA 45-6/L 50 cms.	Needle	90.03	9.06	0.04	0.12	0.00	0.07	0.27	0.09	0.11 +	ti	0.01 +	0.20	------	
DAMARC	24	ABSM/CA 24/L Fisure zone at 'j' rock	Needle	95.83	3.78	0.00	0.14	0.01	0.06	0.05	0.07	0.00	0.00	0.04	0.02	------	
DAMARC	25	ABSM/CA 24-27/L Fisure zone at 'j' rock	Needle	93.67	4.98	0.08	0.77	0.01	0.02	0.00	0.07 +	0.14 +	ti	0.01 +	0.30	------	
DAMARC	26	ABSM/CA 24/ Tall wall / Talayotic.	Frag.	98.81	0.00	0.05	0.32	ti	0.02	0.32	0.04	0.00	0.03	0.00	0.43	------	
DAMARC	27*	ABSM/CA 33/L 40 cms.	Tratinabulum	99.83	0.07	0.00	0.05	0.00	0.00	0.00	0.00	0.00	0.03	0.00	0.02	------	*Corrosion product only.
DAMARC	28	SON PUIG /PRT / Inferior Level	Laurel leaf Dagger	97.00	2.66	0.0	ti	ti	0.02	0.01	0.11	ti	0.00	0.06	0.13	------	
DAMARC	29	LLUCHMAJOR	Tratinabulum	99.37	0.03	0.05	0.09	0.01	0.01	0.09	0.04	0.05 +	0.03	0.00	0.23	------	
DAMARC	30	ABSM /CA 33/L 40 CMS.	Tratinabulum	92.08	0.78	0.18	6.50	ti	0.05	0.00	0.10	0.04	0.02 +	0.00	0.25	------	
DAMARC	31	ABSM / /L	Tratinabulum	94.69	4.96	ti	0.19	ti	0.08	ti	0.00	0.00	0.00	0.00	0.08	------	
DAMARC	32*	ABSM / /L	Sword	0.00	Solo Cu	0.00	0.00	0.00	0.00	0.00	0.00	0.00	0.00	0.00	0.00	------	*Corrosion product only.
DAMARC	33	ABSM /CA 21/ L	Stapte ?	93.94	5.55	0.07	0.03	ti	0.08	0.20	0.07	ti	ti	0.01	0.05	------	
DAMARC	34	ABSM / /L	Strap	89.01	6.65	0.05	3.31	0.02	0.05	0.42	0.11	0.08	0.00	0.00	0.30	------	
DAMARC	35	ABSM / /L	Bull's horn	92.16	6.97	0.08	0.26	0.01	0.08	0.01	0.11	0.13	0.02 +	0.02	0.15	------	
DAMARC	36	ABSM / /L 40 CMS.	Chain track	96.71	2.97	0.07	0.02	0.01	0.06	ti	0.04 +	ti	0.00	0.01 +	0.11	------	
DAMARC	37	ABSM / /L	Tratinabulum	87.62	7.43	0.06	4.41	ti	0.06	0.04	0.04	0.21 +	0.00	0.03 +	0.30	------	
DAMARC	38	ABSM / /L	Knife blade	85.79	13.45	0.1	0.19	ti	0.06	ti	0.18	0.05 +	0.02 +	0.02 +	0.13	------	
DAMARC	39*	ABSM /CA 21/ L	Brooch	98.96	0.31	0.08	0.35	0.00	0.00	0.00	0.14	0.09	0.00	0.01	0.14	------	*Corrosion product only.
DAMARC	40*	ABSM / /L	Spear head	93.29	5.84	0.07	0.55	0.01	0.00	0.01	0.00	0.00	0.02	0.00	0.21	------	*Corrosion product only.
DAMARC	41	ABSM /CA 22/L	Spiral	90.06	8.64	0.05	0.77	0.00	0.08	0.00	0.05	0.08	0.02	0.01 +	0.24	------	
DAMARC	42	ABSM /CA 26-7/L	Spiral	91.72	7.16	0.08	0.52	0.01	0.06	ti	0.02 +	0.00	0.02	0.06	0.35	------	
DAMARC	43	ABSM /CA /L	Disc Tratinabulum	82.89	11.04	0.03	5.40	0.01	0.08	ti	0.07	0.04 +	0.00	0.05 +	0.39	------	
DAMARC	44	ABSM /CA /L	Pin	86.64	12.24	0.64	0.03	0.01	0.08	0.00	0.09	ti	0.00	0.00	0.27	------	
DAMARC	45	SFO/ OS.81/Gred 2 G3	Metal frac.	91.66	6.45	ti	0.30	ti	0.04	0.00	0.09	0.05 +	ti	0.03	1.38	Zn rgoh?	

www.ingramcontent.com/pod-product-compliance
Lightning Source LLC
LaVergne TN
LVHW070046021026
849160LV00027B/1634